T5-AXJ-733

REVEALING THE UNIVERSE

REVEALING THE UNIVERSE

PREDICTION AND PROOF IN ASTRONOMY

EDITED BY
JAMES CORNELL AND ALAN P. LIGHTMAN

The MIT Press
Cambridge, Massachusetts, and London, England

This book was set in VIP Baskerville by DEKR Corporation and printed and bound by Halliday Lithograph in the United States of America.

Library of Congress Cataloging in Publication Data

Main entry under title:
Revealing the Universe.

Bibliography: p.
Includes index.
1. Astronomy. 2. Astrophysics. I. Cornell, James. II. Lightman, Alan P., 1948–
QB43.2.P73 520 81-14301
ISBN 0-262-03080-2 AACR2

CONTENTS

PREFACE

We all have some experience with theory and observation in science. The weather forecaster predicts a sunny day, so we plan a picnic. Early that morning we take a look outside and find it raining. To make his prediction, the forecaster and his research team had to take into account, among other things, the location of high- and low-pressure zones and the dynamics of air masses at various temperatures. His announcement was largely theoretical; your skeptical glance out of the window is observational.

The relationship between theory and observation in science is both complex and symbiotic, and it is one little understood or appreciated by laymen or in some cases even scientists. In a fundamental way observation is the more vital partner in the relationship. Without observations, theory can have nothing to say about what actually does exist; it can only suggest the multitudinous possibilities for what might exist. Interestingly enough, observations, too, require theories. Theory builds the conceptual framework that gives meaning to observational programs. Indeed for all but the most rudimentary experiments this framework is a prerequisite for conceiving a significant and logical observing plan. In the most extreme case, if there were no theory, then all observational data would be just a jumble of facts with no understanding. For example, ancient astronomers gave names to the stars and constellations, but it is the theoretical astrophysicist who can associate the incoming radiation from these objects with their corresponding properties of density, temperature, and composition.

To be sure, then, there are intrinsic differences in the theoretical and observational approaches to a given problem. But in the actual practice of science these distinctions often become blurred, and the two approaches are combined in varying degrees. Biology, for example, deals with relatively complicated and interrelated systems—living organisms—so that it is almost impossible to isolate individual processes sufficiently for a theoretical discussion in the abstract. Instead most biologists must be both theorists and experimentalists. Physics, on the other hand, is a much simpler and more basic science, whose very essence is the study of single, elemental processes, like the swing of a pendulum. Almost all physicists are clearly either theorists or experimentalists.

Finally we must remember that science is a human activity. The persuasive force of theoretical or of observational arguments at any given time is partly a function of our biases. Einstein's 1915 theory of gravity, when applied to cosmology, predicted that the universe could not be static; it must be in a state of either expansion or contraction. But this theoretical prediction disturbed Einstein, and he did not believe the result until observational evidence for expansion was discovered nearly a decade later. Ptolemy's elaborate theory of orbits within orbits to explain the observed motion of each planet was far more complicated than required by the observations. Prejudiced by his philosophical insistence on an Earth-centered solar system, Ptolemy refused to take the observations at face value. Centuries later, Copernicus conceived of the Sun-centered hypothesis, and the observed planetary motions suddenly appeared as simple, single orbits about the Sun.

In this book we explore the relationship between theory and observation within the context of several topics of current interest in astronomy and astrophysics. Chapter 1 gives a historical discussion and chapter 8 a discussion of astrophysical problems for the future. In each of the remaining chapters, with the exception of chapter 6, the same topic is addressed by two scientists, one predominantly a theorist and the other predominantly an experimentalist. Each contributor is an expert in the particular subject, and each views the historical development of the topic and the complementing (and confounding) roles of theory and observation in his own way.

The sometimes divergent, sometimes overlapping viewpoints may be surprising to some readers. Science emerges here not as a neat, systematic progression from theory to observation to testing to solution but rather as a complex and complicated process in which theories and assumptions may lead into blind alleys, observational evidence may challenge theoretical biases, and new discoveries of unimagined phenomena may require entirely new theoretical approaches. In a real sense, then, this is a most honest look at how science and scientists really work. Since science is a human enterprise, it is subject to human failings, prejudices, and mistakes. Yet despite its shortcomings, science demonstrates the power of the human mind to perceive, sort, and select from a wealth of data and a variety of explanations the precise information needed to understand the principles underlying the nature of the physical universe.

ACKNOWLEDGMENTS

Since 1976 the Harvard-Smithsonian Center for Astrophysics and the Charles Hayden Planetarium of the Boston Museum of Science have sponsored a series of free lectures in astronomy for the general public, usually given annually in the fall at the museum. The concept of exploring the interrelated and complementing roles that theory and observation play in modern astrophysics grew out of conversations between the two editors of this volume during their search for a novel approach to the 1980 series. The result was *Revealing the Universe*.

The series was designed with an unusual format in which, with the exception of the introductory and summary talks, each night's presentation would be shared by two speakers, one a theorist, the other an experimentalist. Early in the planning we also realized that the subject and its unique approach might make an interesting book for general readers, and the lecturers were asked to prepare two versions of their talks: the first for the lecture hall and the second for possible inclusion in such a book. The chapters thus correspond to the lectures, roughly in the order they were presented but modified and expanded slightly. The one exception is the chapter by Walter Lewin, which incorporates material presented both by him and by his lecture partner, Paul C. Joss.

The editors wish to thank host John Carr, director of the Hayden Planetarium, who coordinated the lecture series and was most supportive of our ambitious break with tradition. The members of his staff, Matt Stein, Ray Crane, Nancy DiCiaccio, and Valerie Wilcox, were also extremely helpful to our speakers throughout the series. The public lectures themselves were made possible through the generous support of the Lowell Institute.

Assistance was provided by many people at the Smithsonian Astrophysical Observatory, including Earle Hopkins and John Hamwey, who helped in the production and distribution of the original lecture materials. Joseph Singarella and Charles Hanson later provided illustration and photographic services during the preparation of the written chapters. Mary Juliano and Gerda Schrauwen retyped most of the individual chapters, and Anne Omundsen did the final copyediting before the manuscript left the observatory. John Harris of the observatory's Contracts and Grants Office provided valuable advice and counsel at all stages of the project.

CHAPTER 1

OWEN GINGERICH

THE HISTORICAL TENSION BETWEEN ASTRONOMICAL THEORY AND OBSERVATION

Science, one of my wise natural-philosopher friends is wont to say, is not a noun at all; it is a verb. Not grammatically true, perhaps, but the bald claim does say something about the nature of the complex enterprise we call science. Science is not a collection of facts, nor is it a block of specific subjects dealing with the natural world. Instead it is a way of discovering what the natural world is about by conjecturing theories and by subjecting these theories to the test of experiment and observation.

Between the abstract world of theory and the real world of observation there exists a continuous tension and sometimes outright conflict. Why? Because the nexus between theory and observation can be confused by error, and judging whether any disagreement is due to the data or the theory is one of the most difficult tasks confronting a scientist. It is all too easy for the theorist to discount the observations, but there have been glorious moments in the history of science when the theorist's skepticism has been vindicated.

One of Einstein's students has recounted an incident that took place in 1919, just after the British expeditions had returned from viewing the May 29 eclipse and attempting to confirm the relativistic prediction of the bending of starlight around the sun: "Once when I was with Einstein in order to read with him a work that contained many objections against his theory . . . he suddenly interrupted the discussion of the book, reached for a telegram [figure 1] that was lying on the windowsill, and handed it to me with the words, 'Here, perhaps this will interest you.' It was Eddington's cable with the results of measurement of the eclipse expedition. When I was giving expression to my joy that the results coincided with his calculations, he said quite unremoved, 'But I knew that the theory was correct.' And when I asked, 'What if there had been no confirmation of prediction,' he countered, 'Then I would have been sorry for the dear Lord—the theory *is* correct.'"[1]

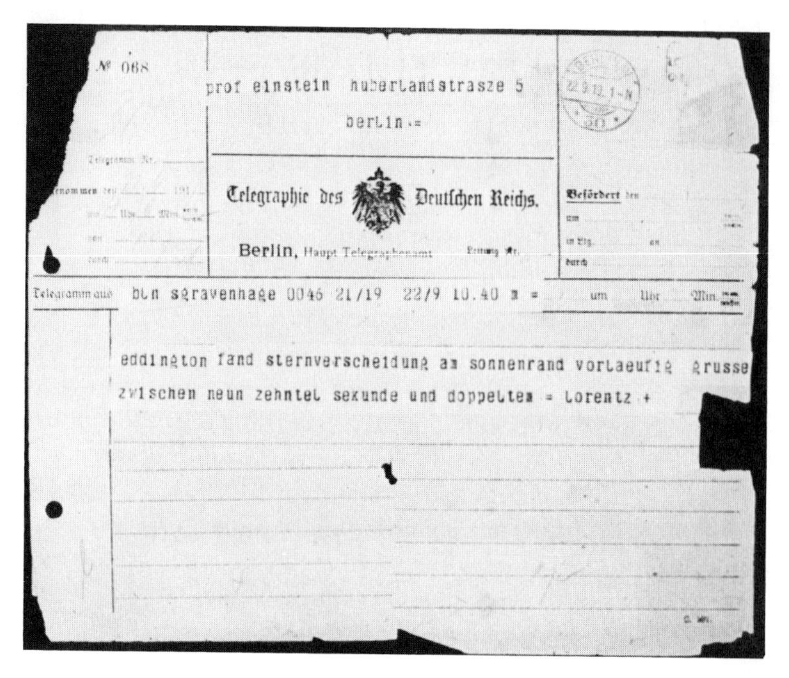

№ 068

prof einstein haberlandstrasze 5

berlin =

Telegraphie des Deutschen Reichs.

Berlin, Haupt-Telegraphenamt

Telegramm aus bln sgravenhage 0046 21/19 22/9 10.40 =

eddington fand sternverscheidung an sonnenrand vorlaeufig grusse zwischen neun zehntel sekunde und doppeltes = lorentz +

Figure 1
The original telegram to Einstein concerning Eddington's successful observation of the bending of starlight near the Sun in the eclipse of May 29, 1919. The telegram was actually sent by the Dutch physicist H. A. Lorentz and states that "Eddington has found a stellar deflection at the solar limb provisionally between 0.9 seconds of arc and twice that." Einstein's prediction was 1″.74. (Copyright Museum Boerhaave, Leiden; used with permission)

All great scientists, both theorists and experimentalists, have had a powerful feeling for the underlying harmony and beauty of the universe. From prehistoric times men and women have learned about our world empirically by doing things: planting and harvesting, spinning and weaving, mining and smelting. But when it came time to build an idealized view of what the universe was ultimately like, the great natural philosophers of antiquity often built their world views on aesthetic principles, relegating observational constraints to second place. Witness Aristotle's arguments in *De coelo* concerning the shape of the Earth. Ordinary sense experience taught that the world was flat, but, moving to a higher level of abstraction, Aristotle argued that "its shape *must* be spherical. For every one of its parts has weight until it reaches the center, and thus when a smaller part is pressed on, it is packed close to and combines with the others until they reach the center. . . . If the Earth has come into being, this must have been the manner of its generation, and it must have grown into a sphere; if, on the other hand, it is ungenerated and everlasting, it must be the same as if it had so developed. Either then it *is* spherical or at least it is natural to be so."[2]

Having given his theoretical argument, Aristotle adds almost as an afterthought that both the Earth's shadow during lunar eclipses and the tales of travelers who saw the northern circumpolar stars drop closer to the horizon as they journeyed southward give evidence for the spherical shape of the Earth. Even today the theoretical argument carries the most power: astronomers believe that no nonrotating body larger than several kilometers in diameter can withstand the gravitational forces that would pull it (roughly) into a sphere.

The case of Claudius Ptolemy (figure 2), the great Alexandrian astronomer who lived six centuries after Aristotle, provides a particularly instructive example about the interaction of theory and observation. Since the easy availability of high-speed computers, it has been possible to calculate planetary positions for early times and hence to check the accuracy of the observations reported in Ptolemy's major treatise, the so-called *Almagest* (in Arabic "the greatest"). I began to do this in the mid-

Figure 2
Claudius Ptolemy, woodcarving in the Ulm Cathedral by Jörg Syrlin the Elder, 1469–1474.

1960s. Quite independently the geophysicist R. R. Newton also began to examine these observations, and we have both noticed something very curious. Ptolemy's observations of the planets are rather erratic, but nevertheless they agree almost perfectly with the positions predicted by his theory. Newton considers this an outrageous state of affairs, and he has concluded that Ptolemy cheated, making up the reported observations to verify a theory that he stole from somewhere else.

I am quite uncomfortable with these accusations. Ptolemy's treatise shows us, for the first time in history, how a numerical theory can be constructed from specific data, and as such it provided the foundation for mathematical astronomy to and through the work of Copernicus. Furthermore Ptolemy's parameters (especially for the difficult cases of the Moon and Mars) are astonishingly accurate, much better than the individual observations reported in the *Almagest.* I am convinced that Ptolemy could not have obtained such a comparatively accurate representation of planetary motion with only the data presented in his book, and therefore it appears that the *Almagest* is written as a paradigmatic text and not as a contemporary research paper with a full explanation of the basic data. Like a modern textbook writer, whose numerical problems often come out even for pedagogical simplicity, Ptolemy probably wished to show readers how it could in principle be done, not how it was actually done.

By the time Ptolemy was ready to present his material, he probably had a good deal more faith in the overall efficacy of his theory than in the individual error-marred observations of his day. Since he did not want to perpetrate any erroneous observations on his unsuspecting students, he may have used his trustworthy theory to remove the observational error from each of his data points. Then each observation would agree with the positions calculated from his table, as indeed they do. Unlike R. R. Newton, I am unwilling to condemn Ptolemy as a criminal if, in fact, he has quietly deleted the observational error. For me, the episode gives some insight into the nature and expectations of ancient science rather than into the fraudulent character of Claudius Ptolemy.[3]

Some commentators seem to have completely misunderstood Ptolemy's role in the progress of science, somehow believing that his treatment of data and his success prevented an earlier acceptance of a heliocentric astronomy. Such a statement appeared in the third episode of the "Cosmos" television series, where Carl Sagan declared that the church and the Ptolemaic system held back the progress of astronomy for over a millennium. The clear counterexample is Chinese astronomy: for better or worse, China had neither Christianity nor the Ptolemaic system, yet its astronomy remained far more primitive than in the West.

In fact, when Copernicus finally appeared, he founded his heliocentric astronomy on precisely the observations of Ptolemy—and poor as those were, they were considerably better than some that Copernicus incorporated from his own day. Ironically Copernicus was quite aware of shortcomings of Ptolemy's planetary predictions, but he never capitalized on this knowledge. Shortly after he had returned to his native Poland from graduate studies in Italy, Copernicus observed the rare multiple conjunctions of Mars, Jupiter, and Saturn that took place in 1503 and 1504. It did not require sophisticated instruments to find the moments when the closest approaches took place, and with these observations Copernicus promptly found that both Mars and Saturn were more than a degree from the places Ptolemy had predicted. The Polish astronomer recorded the fact in his notebook, but he never mentioned it in print. The reason? The move to the heliocentric cosmology took place not on account of observational evidence but on theoretical grounds that can perhaps best be called aesthetic. Because the underlying observations were so similar for both Ptolemy and Copernicus, the predictions of planetary positions were only slightly improved by the tables in Copernicus's treatise.

As Galileo later remarked, he could not admire enough those who had adopted the heliocentric viewpoint despite the evidence of their senses. The simple fact was that Copernicus had no proof that the Earth went around the Sun. His theory had advanced a world view one level of abstraction higher than before, a view that was in some respects powerfully more beautiful and coherent but one that lacked any decisive observational underpinnings. Even Galileo, who achieved enviable polemical gains for heliocentrism through his telescopic discoveries, could do no more than argue for the probability of the new doctrine (figure 3). Cardinal Bellarmine was on solid logical grounds when he wrote to Foscarini, "It appears to me that you and Signor Galileo did prudently to content yourselves with speaking hypothetically. . . . If there were a true demonstration that

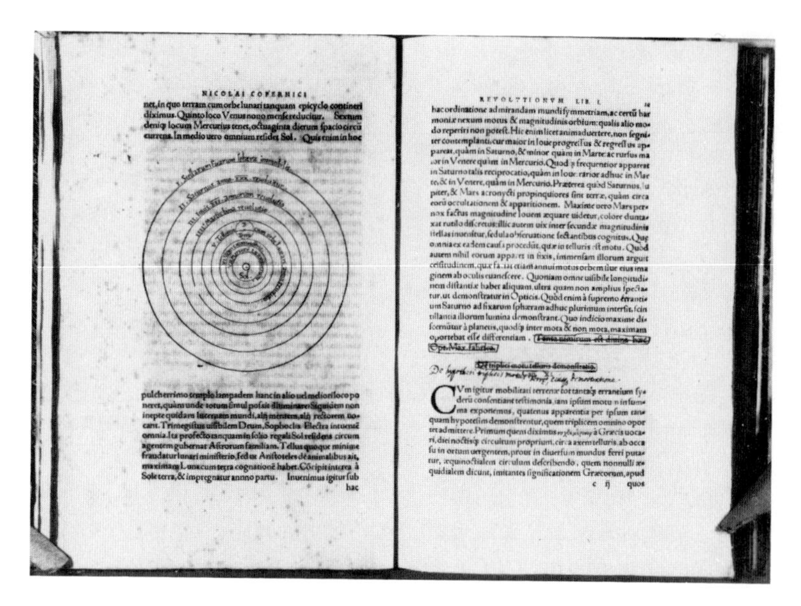

Figure 3
The Copernican system, shown in Galileo's copy of his second edition of *De revolutionibus* (Basel, 1566). The censorship required by the Inquisition has been carried out by Galileo in his own hand on the right-hand side: the chapter heading "On the explication of the three-fold motion of the earth" has been changed to read "On the hypothesis of the three-fold motion of the earth and its explication." (Permission of the Biblioteca Nazionale Centrale, Florence)

the sun was in the center of the universe and the earth in the third sphere, then it would be necessary to use careful consideration in explaining the Scriptures that seemed contrary. But I do not think there is any such demonstration, as none has been shown to me. To demonstrate that the appearances are saved by assuming the sun at the center and the earth in the heavens is not the same thing as to demonstrate that in fact the sun is in the center and the earth in the heavens."[4]

What, then, were the aesthetic reasons that led Copernicus to his radical, Sun-centered cosmology? The challenging peculiarity of the planets' apparent motions among the fixed stars is their occasional retrograde or westward movement. Ptolemy accounted for these observations by supposing that the direct motions in two circles combined to produce the retrogression. Indeed this is what the Copernican system also requires: the direct motion of the Earth in its orbit and of each planet in its orbit. The difference between the systems is that for Ptolemy the two-circle combination is independent and arbitrary for each planet, whereas for Copernicus the Earth's orbit invariably takes the place of one of the two circles. For Ptolemy, the ratio of the sizes of the two circles was fixed by observational constraints, but there was no required connection between one planet and another. For Copernicus, once the size of the Earth's orbit is fixed, the ratio of each planetary orbit is locked with respect to it, and thus the whole system was linked together, "as if by a golden chain" in the picturesque words of Copernicus's only disciple, Rheticus. This, then, is the essence of Copernicus's aesthetic achievement as we admire it today.

But in Copernicus's century there was more. Copernicus was not only overwhelmingly convinced about this heliocentric cosmology—so much that he was willing to abandon the time-honored Aristotelian physics with no replacement in sight—but he also felt strongly about the ancient Pythagorean precept that the eternal celestial motions must be composed of pure circles. Hence Copernicus went to great lengths to purge Ptolemaic astronomy of devices that, in his opinion, violated the uniform circular motion rule. And in the sixteenth century Copernicus was admired more for this accomplishment than for stopping the Sun and moving the Earth.

Thus Copernicus's research program had two goals: to demonstrate a heliocentric cosmology and to reestablish a system with uniform circular motions. I mention this to show not only how right theoretical intuition can be but also how wrong it can be. It remained for Kepler, several generations later, to put the kibosh on circular orbits by establishing their elliptical form.

Some astronomical data from an entirely different century are instructive here.[5] Figure 4 shows the relative positions of the bright star Vega as measured in 1836–1837 by Friedrich Georg Wilhelm Struve at the Dorpat Observatory in Russia. These data show the first tenuously correct measurements of stellar parallax, that is, the annual shift of a star's position arising from the yearly revolution of the Earth in its orbit. We might say that it is the first physical proof of the Copernican system, but in 1837 no one took these data very seriously. One of my students once looked at this graph and asked, "Why is there so much error?" If you notice the units, you will see that Struve was measuring to a few hundredths of a second of arc—and a second of arc, one part in thirty-six hundred of a degree, is the angular size of a dime seen at the distance of a kilometer. The

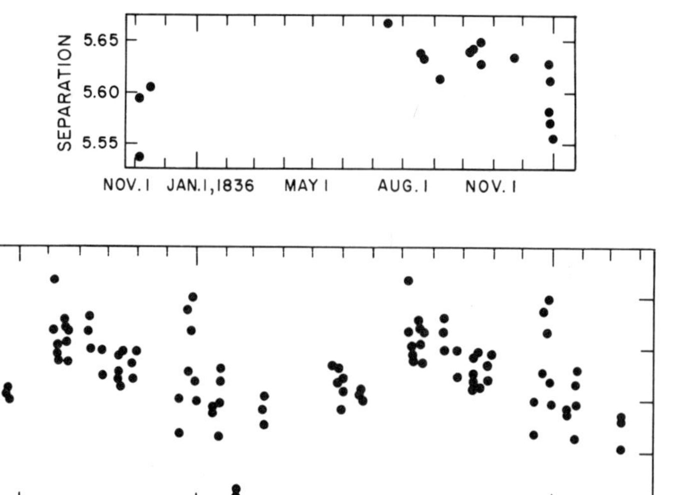

Figure 4
Friedrich Georg Wilhelm Struve measured the position of Vega with respect to a star lying nearby (but which was actually so far away that its annual parallax was not measurable). The separations are in seconds of arc. The upper series shows the controversial observations from November 1835 to the end of 1836; the lower series, published after Bessel had announced his own measurements of 61 Cygni, clearly shows the cyclic annual shift in position. In the lower series, the measurements for February 1837 through January 1838 have been repeated in order to show the sinusoidal variation more clearly. Can such variation be convincingly detected from the upper series?

average angular size of Vega was probably three seconds of arc. Perhaps the proper question is, How could Struve have done it so well?

Now let me return to Kepler, because his problem was something like Struve's, that is, how to sort out the shape of the orbit from noisy data. The standard story is that Tycho Brahe's observations, inherited by Kepler, were just good enough so that finding the elliptical shape of Mars's path was inevitable. The truth is that Kepler was looking for a most subtle effect, and he stumbled onto the ellipse only because it was a relatively easy curve to handle. But Kepler remained skeptical until he was finally able to jinn up a theoretical-physical explanation as to why the planet should go in an elliptical orbit. Part of his explanation was that magnetic emanations from the Sun pushed the planets around, as well as alternatively attracting and repelling them into the elliptical paths. And this in turn presupposed that the Sun itself was rotating, and rotating faster than the orbital period of Mercury. Although Kepler's magnetical explanation was entirely wrong, it contained the essential idea that some physical property of the Sun governed the planets in a fashion that could be expressed mathematically. Part of Kepler's greatness was his insistence that such explanations must exist.

Not long thereafter Galileo, with the newly invented telescope, discovered sunspots and, from this, the rotation of the Sun. Kepler was understandably miffed when Galileo gave him no credit for a theoretical prediction of the solar rotation. But should he have this credit when the theoretical foundation for his prediction was false?

Indeed Isaac Newton was very stingy with credit to Kepler even for the elliptical planetary orbits, barely mentioning his name and saying elsewhere that Kepler had "guessed" the shape but that he, Isaac Newton, had shown with his theory of universal gravitation why the orbits had to be elliptical. Newton's contemporaries were more generous to Kepler, and, when Edmond Halley came to review Newton's *Principia,* he wrote that the opening propositions were "found to agree with the *Phenomena* of the Celestial Motions, as discovered by the great Sagacity and Diligence of *Kepler.*"

Newton, like many other great theorists, had immense confidence in the efficacy of his theory. It is almost amusing to see him writing to the Astronomer Royal, John Flamsteed, to thank him for a set of lunar observations: "I am of opinion that for your Observations to come abroad with [my] Theory . . . would be much more for their advantage and your reputation then [*sic*] to keep them private till you dye or publish them without such a Theory to recommend them. For such a Theory will be a demonstration of their exactness and make you readily acknowledged the Exactest Observer that has hitherto appeared in the world."[6]

One of the fundamental causes of the tension between theory and observation is the ambiguous borderline between the two; that is, observations are perceived through a particular theoretical framework, and they are rarely if ever theory free. Let us look at two mental inventions of the seventeenth century, the satellites of Jupiter and the rings of Saturn. "Wait a minute!" you might exclaim. "How can pure observations be inventions?"

Galileo did not observe the satellites of Jupiter. What he observed were small points of light accompanying the planet and changing their relative positions from night to night. It was his interpretation that those brilliant specks oscillated continuously from one side of the planet to the other and that this oscillation could be understood theoretically as the consequence of circular motion. The interpretation, of course, assumed that there was also an unobserved component of motion toward and away from the observer. Merely observing the tiny points of light did not guarantee an instant interpretation. Only recently have astronomers noticed that on two occasions in 1613 Galileo actually observed the planet Neptune among the satellites and that he even suspected its motion. Nevertheless his perceptive framework was not sharp enough for him to have followed up on what could have been a truly staggering discovery.

The case of Christiaan Huygens and the rings of Saturn is even clearer. Galileo first observed that Saturn appeared to have "handles," and he was mystified when these appendages later disappeared. Forty years later telescopes were much improved, but when Huygens hit upon the correct interpretation of the "handles" as rings, they were edge-on and invisible. Huygens achieved his solution by theoretical considerations of symmetry and of geometry. Perhaps not entirely convinced by his mental model, Huygens coded his results into a Latin anagram and waited a year until the rings became visible again before announcing their nature publicly (figure 5).

Another facet of the historical tension between theory and observation is the difficulty of recognizing and interpreting ambiguous data. The discovery of stellar parallax provides a particularly clear example. For many years textbooks were almost unanimous in ascribing the priority to the results achieved by the German astronomer Friedrich Wilhelm Bessel in 1838. However, in the wave of nationalism that swept the Soviet Union in the days of the Cold War, the Soviets laid claim to a number of innovations previously associated with Western scientists and inventors. Among these contentions was the priority for the detection of stellar parallax, by F. G. W. Struve at the Dorpat Observatory in 1837. Subsequently Struve's great-grandson, Otto Struve, examined the printed record.[7]

Otto Struve's analysis of the situation may be characterized as one involving a confidence quotient. Astronomers had long been searching for this tiny displacement of a star's position

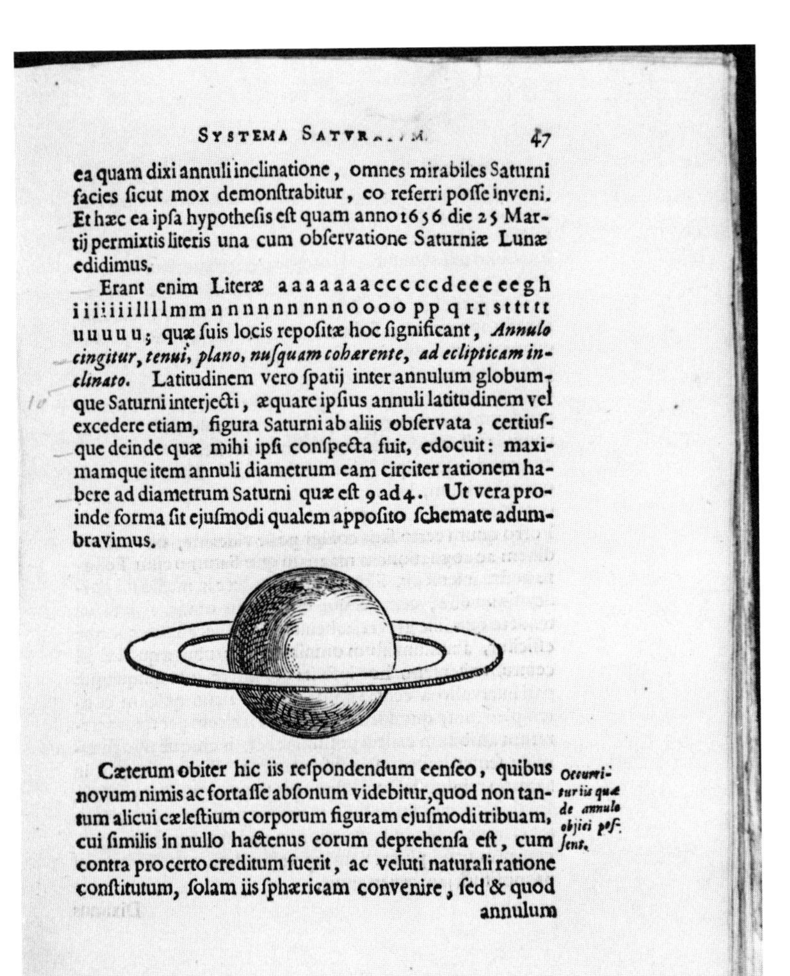

SYSTEMA SATVR...M. 47

ea quam dixi annuli inclinatione, omnes mirabiles Saturni facies ſicut mox demonſtrabitur, eo referri poſſe inveni. Et hæc ea ipſa hypotheſis eſt quam anno 1656 die 25 Martij permixtis literis una cum obſervatione Saturniæ Lunæ edidimus.

Erant enim Literæ aaaaaaacccccdeeeeegh iiiiiiillllmmnnnnnnnnnoooopp qrrsttttt uuuuu; quæ ſuis locis repoſitæ hoc ſignificant, *Annulo cingitur, tenui, plano, nuſquam cohærente, ad eclipticam inclinato.* Latitudinem vero ſpatij inter annulum globumque Saturni interjecti, æquare ipſius annuli latitudinem vel excedere etiam, figura Saturni ab aliis obſervata, certiuſque deinde quæ mihi ipſi conſpecta fuit, edocuit: maximamque item annuli diametrum eam circiter rationem habere ad diametrum Saturni quæ eſt 9 ad 4. Ut vera proinde forma ſit ejuſmodi qualem appoſito ſchemate adumbravimus.

Occurritur iis quæ de annulo objici poſſent.

Cæterum obiter hic iis reſpondendum cenſeo, quibus novum nimis ac fortaſſe abſonum videbitur, quod non tantum alicui cæleſtium corporum figuram ejuſmodi tribuam, cui ſimilis in nullo hactenus eorum deprehenſa eſt, cum contra pro certo creditum fuerit, ac veluti naturali ratione conſtitutum, ſolam iis ſphæricam convenire, ſed & quod annulum

Figure 5
On this page of his *Systema Saturnum* Huygens deciphers his anagram (given at the beginning of the paragraph) to reveal that Saturn is "girdled by a thin, flat ring, nowhere touching it, and inclined to the ecliptic." (Permission of Houghton Library, Harvard University)

with respect to much more distant stars that was predicted as a result of the Earth's motion in the accepted Copernican system. Bradley, for example, had discovered in 1725 a corollary proof of the Copernican system, the aberration of starlight, while searching in vain for stellar parallax. (The aberration of starlight is the change in angle of rays arising from the *velocity* of the Earth, similar to the apparent change in the direction of falling raindrops when we run through the rain. This differs from the parallax effect, which arises from the changing *position* of the Earth.) By the early 1800s several announced detections of this subtle phenomenon had been proved false. Hence when F. G. W. Struve found a marginal effect nearly buried in the scatter of his observations of the star Vega (shown in figure 4), astronomers could not be sure whether he had measured the elusive displacement. Only in retrospect, after the far more convincing demonstration of the annual parallactic displacement of 61 Cygni by Bessel, could astronomers be sure what Struve had found. Such are the ambiguities of the data at the cutting edge of research and the ambiguities of assigning priorities.

Indeed the tension between theory and observation would entirely disappear if the observations were sufficiently complete and unambiguous, but then we would not be considering the frontiers of knowledge. Precisely because the research reported in this book is frequently at the cutting edges, such ambiguities repeatedly appear. For example, how seriously can we take the observation of the ultraviolet emission from deuterium, which seems to tell us about the density of the universe in its initial stages? Are the preliminary observations of the mass of neutrinos persuasive? If so, do they give us a way out concerning the missing neutrinos in the "underground astronomy" experiment in the Homestake Gold Mine? (See chapter 8, George Field.) Are the radial velocity observations of Cygnus X-1 convincing enough to believe that a black hole has been found in a binary star system? (See chapter 5, Alan Lightman.) And will these queries lead us to contradictory answers about whether the universe will eventually collapse back in upon itself? (See chapter 7, John Huchra.)

One of the most amazing aspects of science is that questions are often answered in remarkably unexpected ways, and these are generally associated with fresh data. To a certain extent these new data come to hand through advances in technology, and often the technological innovators become the foremost observers. Photography, spectroscopy, and the building of large reflectors each brought floods of new data that have transformed the face of astronomy. Radio telescopes, infrared detectors, and x-ray cameras currently are the tools broadening our astronomical horizons. But it is not only a qualitatively but a quantitatively wider access to data that is important for the advance of science. One might well argue that Ptolemy's achievement was at least in part owing to the resources provided by the great library at Alexandria and that Copernicus reached his conclusions when he did because the increasingly widespread use of printing had given him an easier access to sources than his predecessors enjoyed. Kepler could go much further than Copernicus because of the feast of observations provided by Tycho, and Galileo could defend the heliocentric viewpoint more convincingly than Copernicus because of the astonishing new data from his telescope. Today, the computer and electronic multichannel observing devices give both theorists and observers incredibly richer resources than the astronomers of the previous generation.

Whenever science is on the frontier of knowledge, an essential tension will exist between theory and observation. And whenever this essential tension decays, science as a creative human enterprise also wanes. The forward march of science has not always been uniform, as its progress during the Middle Ages has shown all too clearly. When the structure of science and society is finely tuned and working optimally, there is a constant interaction between theory and observation, but this interaction is mediated by the efficiency of our data-gathering facilities.

Humanity's ability to glean and preserve new data precipitously declined with the decay of the Alexandrian library. In the energy-rich and technologically sophisticated society in sixteenth-century Tuscany, fertile conditions arose on which was built a flourishing culture and science, but these conditions rapidly declined with warfare and with the censorship of books. The torch of science quickly passed to northern Europe and England. Late in the last century, preeminence in astronomy moved to the United States. The reason is not hard to find: through its wealth and philanthropy, and thereby through its ever-larger telescopes, America became the largest supplier of astronomical observations. It is true that brilliant theoretical work and observations continued in Europe, but by any reasonable criterion (such as the indexes in histories of astronomy not published in the United States or the gold medals awarded by the Royal Astronomical Society) the centrum had clearly moved to America.[8] Throughout the twentieth century patterns of funding have evolved—government grants have replaced the largesse of industrial magnates—but there has continued an unabated flow of data from strange and previously unexplored worlds: Mercury and Io, the molecular complexes of Orion, the nucleus of the Milky Way, quasars, and the primeval fireball.

To maintain the forward march of science and to advance our understanding of the material universe in which we live require the ongoing engagement of theorists, observers, and theorist-observers. But the foundation of all their endeavors is the data base, and as we move toward the twenty-first century, we risk intellectual stagnation if we fail to maintain not only healthy houses of wisdom for preserving the data already in hand but also the invaluable tools of exploration for gathering new data about our universe. And if science continues to cut new paths into the unknown, creative tension between theory and observation will remain at the frontier.

Notes

1. Quoted by Gerald Holton, *Thematic Origins of Scientific Thought* (Cambridge, 1973), pp. 236–237, from a manuscript by Ilse Rosenthal-Schneider.

2. Paraphrased from W. K. C. Guthrie's translation of *On the Heavens* (Cambridge: Leob Classical Library, 1939) and appearing in M. K. Munitz, ed., *Theories of the Universe* (New York, 1965), pp. 98–99.

3. For an exchange of views on this subject see my article "Was Ptolemy A Fraud?" in the *Quarterly Journal of the Royal Astronomical Society* 21 (1980): 253–266, R. R. Newton's reply (ibid. 21 [1980]: 388–389), and my further remarks "Ptolemy Revisited" (ibid. 22 [1981]: 40–44).

4. Translated by Stillman Drake, in his *Discoveries and Opinions of Galileo* (Garden City, N.Y., 1957), pp. 162–164.

5. From Otto Struve, "The First Determinations of Stellar Parallax," *Sky and Telescope* 16 (1956):9–12, 69–72.

6. Newton to Flamsteed, 16 February 1694/5, in *Correspondence of Isaac Newton* (Cambridge, England, 1967), 4:87.

7. Struve, "First Determinations."

8. Stephen G. Brush, "The Rise of Astronomy in America," *American Studies* 20 (1979):41–67.

CHAPTER 2

EINSTEIN'S PERCEPTIONS OF SPACE AND TIME

KENNETH BRECHER

NEWTON, EINSTEIN, AND GRAVITY

I want to know how God created this world. I am not interested in this or that phenomenon, in the spectrum of this or that element. I want to know his thoughts, the rest are details.

—A. Einstein[1]

Newtonian gravitation. For nearly three centuries, these words stood for the epitome of human intellectual achievement. More than a theory of gravitation, a physical picture, or a mathematical structure, the idea conveyed by these words represented the very model of human understanding of nature. It pointed the way for a later understanding of fluids, optics, electromagnetism—in fact, all of classical physics. And it was wrong.

By the turn of the twentieth century Newtonian gravitation had become the most accurate representation of any process in nature, for it could accurately predict and elegantly describe the motions of the planets (with one small exception), comets, and even distant stars. Yet at its deepest roots, in its very essence (not its predictions) and in its philosophical underpinning, the concept bore only slightly more connection with nature's reality than did Ptolemy's awkward epicyclic views of the heavens.

Indeed until Einstein began to think about gravitation, no one even suspected that an entirely new world view was necessary to deal with this very central part of physics. Why had no one suspected the faults in Newtonian thinking? And what failed predictions, what obvious inconsistencies and confusions, could be found within Newtonian gravitation that might lead directly to a new *Weltbild*? The answer is that there were precious few clues to a failure of the classical view; rather it was primarily Einstein's philosophical perceptions of space and time that pointed the way to the new, and ultimately more successful, theory of gravitation, the General Theory of Relativity.

This story has no real beginning, so I will arbitrarily choose 1905 as the start, when Albert Einstein, at the age of twenty-six, proposed his Special Theory of Relativity. (One could equally well begin a decade earlier when Einstein began to ponder what it would be like to travel along at the speed of light and view a moving beam of light, or even half a century earlier, when Georg F. B. Riemann began to wonder whether space itself might be a by-product of gravitation.)

Einstein's theory was constructed in response to the difficulty in reconciling James Clerk Maxwell's triumphal formulation of electromagnetic theory with Newton's mechanics and its associated concept of absolute space and time. After a ten-year search Einstein had found that these two pictures could be reconciled only if time—previously viewed as uniform and evenly flowing for all observers—were a relative concept, dependent on the motion of the observer.

In a sense, another by-product (although it could as well be taken as a basis) of the Special Theory of Relativity (which I will refer to as Special Relativity) is that the speed of light must be constant and independent of the motion of its source or of any observer. In addition, it appeared that this meant that nothing could move faster than the speed of light. Newtonian gravitation is an action-at-a-distance theory, where the effects of a gravitational field are felt instantaneously at arbitrarily great distances from a moving source. Therefore the force of gravity, according to Newton, propagates at an infinite speed. It was clear to Einstein that some revision in Newtonian gravitation was in order.

If such a revision was to take place, what guidelines for its development were there? And what signposts pointed to errors in the previous theory that might guide one to the truth? After all the predictions of Newtonian gravitation were highly accurate. In fact, in all applications of the theory, the only discrepancy between prediction and proof had been the small movement in the perihelion of the orbit of the planet Mercury. Planetary orbits are not perfect circles but ellipses, and the perihelion of an orbit is the position of an object's closest approach to the Sun. The movement of Mercury's perihelion was well established by the French astronomer Urbain Jean Joseph Leverrier in 1845 and could be explained within Newtonian gravitation only if the Sun were sufficiently nonspherical or if there existed, for example, a small (until then unseen) planet lying within the orbit of Mercury and perturbing Mercury's orbit. This difficulty was rather little to prompt development of a whole new theory and, to my knowledge, never resulted in the construction of one.

One other small problem remained in Newton's theory, however; and again it was the genius of Einstein not only to recognize it, but also to perceive its importance. Early in the seventeenth century, so the story goes, Galileo leaned over the edge of the Tower of Pisa, dropped two objects made of different materials, and noted, to no one's surprise, that they hit the ground at essentially the same moment. This experiment, implying that all objects fall in a gravitational field in the same way independent of their composition, was subsequently taught for the succeeding three centuries to school children and future physics professors alike, without making much of a ripple. Even in the late nineteenth century, when the Hungarian physicist Baron von Eotvos repeated Galileo's experiment with a precision of about one part in a billion, little seems to have been made of the point. This result, formulated in a statement called the equivalence principle, troubled Einstein, however. Why, he asked, should gravitation act on all objects in the same way? Even more significantly, he began to ponder what this experimental result implied about the fundamental nature of gravitation. Further, the equivalence principle was not an intrinsic property of Newtonian gravitation; rather it had to be added to the theory.

It would be nice for this story if, as in the case of Special Relativity, Einstein was led directly to a triumphal new theory, this time of gravitation. This did not come about quite so simply, and the development of General Relativity was a rather unusual example of the connection between prediction and proof in the search for knowledge of the physical world.

Shortly after the formulation of Special Relativity, Einstein demonstrated the equivalence of mass and energy. Each amount of mass has stored within it an amount of energy equal to the mass multiplied by the speed of light squared: $E = mc^2$. Einstein reasoned that if mass is affected by a gravitational field and mass and energy are equivalent, any type of energy (for example, a light wave) also should be affected by the presence of gravitation. And in 1907 he showed that if light moves radially out of a massive object, the gravitational field of that object will pull on the light waves, so that the light will lose some of its energy and become redder in color. In short gravitation not only slows down projectiles leaving a body such as the Earth, it also does work on light, changing its energy and frequency while leaving the speed of the light constant.

By 1911 Einstein had developed additional ideas on gravitation that he thought might incorporate the main requirements of Special Relativity, as well as the results of the Eotvos experiment. Although not yet a real theory, Einstein's 1911 paper nevertheless made another novel prediction. Light, like any other kind of ponderable matter, must have its motion affected as it passes by a massive object. A light ray from a distant star that grazes the edge of the Sun should be pulled toward the Sun, thus making the apparent angular position of that star move away from the Sun (figure 1). Einstein wrote in 1911, "A ray of light going past the Sun would accordingly undergo reflexion to the amount of $4 \times 10^{-6} = 0.86$ seconds of arc (two ten-thousandths of a degree). The angular distance of the star from the centre of the Sun appears to increase by this amount. As the fixed stars in the parts of the sky near the Sun are visible during the total eclipses of the Sun, this consequence of the theory may be compared with experience. . . . It would be a most desirable thing if astronomers would take up the question here raised. For apart from any theory there is the question whether it is possible with the equipment at present available to detect an influence of gravitational fields on the propagation of light."[2]

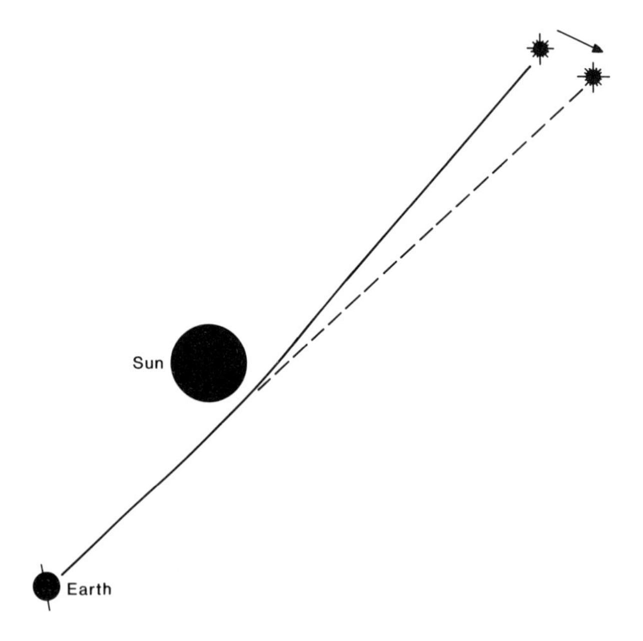

Figure 1
Deflection of light by the Sun.

The last sentence, in addition to displaying Einstein's modesty, indicates his interest in the value of experimental confirmation of his theoretically motivated ideas. As to the astronomers' response to his challenge, one astronomer, Erwin Freundlich of the Royal Observatory in Berlin, was sufficiently impressed by the young man's speculations to plan an expedition to a region of total eclipse of the Sun. Unfortunately for him, but fortunately for Einstein, the next satisfactory location to observe an eclipse was in the Russian Crimea in 1914. A few weeks after Freundlich arrived in Russia for his expedition, World War I broke out, he was arrested by the Russians, and the measurement was not made. (Later Freundlich was exchanged for Russian officers captured by the Germans.)

What was so fortunate about this set of circumstances? As it turns out, Einstein had made a quantitative (although not qualitative) error. As he had not yet worked out the full theory of gravitation, his preliminary calculation gave half the correct value of the deflection of light by the Sun. Of course, that there should be some such observed effect was correct; and after all Einstein had felt this was the main point "apart from any theory." (Somewhat similarly and for not unrelated reasons, the physicist J. von Soldner in 1801 had predicted an angular deflection of 0.86 seconds of arc for starlight by the Sun.)

The year following Freundlich's abortive expedition, Einstein finally completed his theory of gravitation, the General Theory of Relativity (General Relativity). One of the first applications of the new theory was to his previous problem, the deflection of starlight by the Sun. To his surprise he found that the true deflection should be approximately 1.73 seconds of arc, or twice the previously calculated result.

World War I drew to a close and, in the postwar shambles, a few farsighted scientists saw the immense importance of Einstein's work. Among these was the British theoretical physicist and mathematician Arthur Stanley Eddington. Although not an experimentalist by training or vocation, he believed that Einstein's revolutionary views of space and time, and their embodiment in General Relativity, should be tested if at all possible. So in 1919 he set out for the island of Principe off the coast of Africa to observe a very favorable total eclipse of the Sun. The day arrived, as luck would have it rather cloudy, but clear enough to allow Eddington to get one good photograph of stars in the neighborhood of the Sun during the eclipse. Within a few weeks, Eddington had measured the star positions and compared them with a photo of the same star field taken six months earlier. The deflection was there, within an accuracy of 10 percent just as Einstein had predicted it in 1915.

The news swept the globe: Albert Einstein had replaced Sir Isaac Newton as the premier physicist of all time. A war-weary world was treated to the image of a German theoretical physicist's overthrowing the British giant, with the help of another member of the Commonwealth. Viewed more positively, this seemed the triumph of scientific collaboration between the recently opposing nationalities in an otherwise irrational world. Headlines in the *New York Times, Le Monde,* and the *Times of London* all blared forth the news, and the shy, retiring, and previously obscure scientist became the darling of both the scientific and general publics. In an age when mass communications systems were suddenly burgeoning worldwide, Einstein was the first true media personality. (To make a parallel to our own time, Einstein was about as popular as the Beatles, and his popularity and renown were due in no small part to the successful corroboration of his prediction.)

Imagine now that there had been no World War I and that Freundlich had succeeded in making the original measurement. He might well have found that the observed result was twice Einstein's predicted value of 0″.86. How then would people have viewed the theory? Would they have said, Well, the effect is there, but one must now go back and modify the theory to accommodate the result? Or would Freundlich have gone out to another eclipse to discover what had gone wrong with his first measurement? It is difficult to say for sure. But it is almost certain that given human nature, Einstein would not have been instantaneously lifted to world renown. Ultimately he probably would have completed his theory (although one never knows how the first wrong prediction might have affected him), gotten the correct result, and then moved ahead.

At least two other applications of General Relativity shed light on the peculiar way in which theory and observations affect one another in the progress of science and the general acceptance of new ideas by professional scientists and the public alike.

As a work of physics, as an application of mathematics, and as a philosophical approach to geometry, gravitation, space, and time, General Relativity was an absolute triumph of the human mind, but it still had to agree with reality. Another early appli-

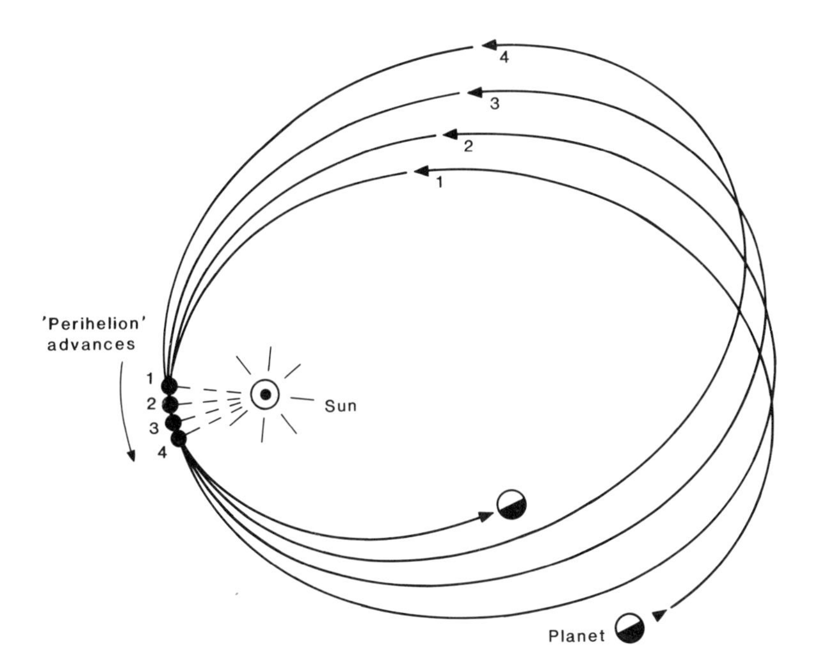

Figure 2
Perihelion advance.

cation of the new theory was to the motion of the planets. When gravity is weak and motions are slow, General Relativity is well approximated by the Newtonian theory of gravity. As the strength of the gravitational field is increased, small deviations between the two theories grow larger. One of the resulting effects is that orbits that have fixed perihelia in Newtonian gravitation should have slowly rotating perihelia in General Relativity (figure 2). The amount of this rotation for the planet Mercury is a mere one-hundredth of a degree per century, an almost infinitesimal effect but one well within the capability of observational planetary orbit studies to discern. Leverrier had first found this discrepancy in 1845, and the problem was still there in 1915. Thus, in retrospect Einstein had solved the only major error in celestial mechanics as a by-product of his theory. Within the observational accuracy of the time, the observation and theory agreed perfectly. This, then, was the second (albeit after the fact) triumph of General Relativity. Still even this success was not one that might catapult General Relativity to a dominant position over the old Newtonian theory. But the final nail in the Newtonian coffin was yet to be put in.

Einstein's first prediction of an observable phenomenon resulting from the merging of Special Relativity with gravity was that of the shift of spectral colors toward the red end of the spectrum (through comparison of solar with terrestrial sources of radiation) by about two parts in a million. (This effect is described in a footnote to his 1911 paper.) Of course, even in the 1980s, the measurement of such a small effect against the turbulent background of the Sun's photosphere is spectacularly difficult. Luckily, however, there are stars in which the effect should be much larger than on the Sun, so a non-solar-system test was suggested after 1919 when confidence in General Relativity was high.

Walter Adams, an astronomer in California, had been interested in the properties of what were then some poorly understood and rather dim stars, among them the binary companion of the brightest star in the sky, Sirius. This star, ten thousand times dimmer than its companion, nonetheless was shown (by application of Newtonian gravitation) to have a mass about the same as that of the Sun (figure 3). Yet Adams had found its temperature to be quite high, perhaps 8000 K. This meant that the star had a small radius, a little larger than that of the Earth. Based on the suspected mass and radius of the star, it should have had a surface gravity field strong enough to give rise to a

Figure 3
Sirius and white dwarf companion at extreme right; three bright points are images of Sirius caused by lens reflections. (Photo from U.S. Naval Observatory)

shift in spectral wavelength of six parts in a hundred thousand. At the request of Eddington, Adams set out to make the enormously difficult measurement: to try to measure the colors of a star ten thousand times dimmer than a companion star only three-thousandths of a degree away. Adams reported his results in a series of papers in 1925.[3] Einstein's theory predicted a red shift of six parts in a hundred thousand, and Adams had found just such an effect.

General Relativity had triumphed once more, and again Einstein was front-page news. The publicity was not so world-sweeping as before, for by now it was already accepted that Einstein had replaced Newton. More important, the Theory of General Relativity was considered true, and its predictions, like those of Newton before, were now simply numbers to be checked in the normal routine operation of science. By 1925, with the confirmation of Newtonian gravitation behind it, plus the revelation of a small but important set of new facts, General Relativity clearly seemed to be the superior theory and the new standard for describing gravity.

In fact, the belief in the absolute validity of General Relativity was so widespread that Adams's stunning confirmation of the theory may have been little more than wishful thinking. Not that the red shift produced by Sirius disagrees with theory; rather both the theoretical prediction (which depends on the assumed mass and radius of Sirius) and the observations by Adams were wrong, but in just the right sense so that they would agree. This was not a result of malicious intent but probably because Adams already knew what he was looking for and found it. Only in 1971, through the work of Jesse Greenstein and his colleagues, was the situation finally resolved.[4] The mass of Sirius had been slightly underestimated and its radius overestimated, so that the predicted red shift should have been about twenty-eight parts in a hundred thousand. This value was indeed measured in the later work, thereby confirming General Relativity with an accuracy of better than 10 percent.

From the mid-1920s until the mid-1960s, General Relativity remained a fixed but somewhat peripheral part of physics. Most physicists had shifted their interests from space-time studies to quantum theory, atomic physics, and nuclear physics, and then after World War II to elementary particle physics. Yet all this time General Relativity remained the accepted theory of gravitation. A few new ideas were added to it, a few new predictions made, including the expansion of the Universe as a whole (see chapter 7, John Huchra and William Press); but as a branch of physics, the study of General Relativity was by and large something of a scientific backwater.

The last twenty years, however, have brought with them a renewed interest in General Relativity, for two somewhat interconnected reasons. The first is technological. More elaborate and accurate instruments now enable us to improve old experiments and conduct new tests of Einstein's theories with ever-greater precision. The second reason has been the application of these new technologies to astronomy. The development of radio, infrared, ultraviolet, x-ray, and gamma-ray astronomy and the creation of larger telescopes both on the ground and in space have led to exciting and profound discoveries about the nature of the universe. And the discovery of powerful astronomical objects such as pulsars and quasars suggests that General Relativity plays a vastly more important role in the cosmos than simply introducing small corrections to Newtonian gravitation.

In the modern era, then, the story of General Relativity divides itself naturally into two parts. First are the more precise tests of General Relativity within the solar system, where gravity is relatively weak. And second are the tests involving strong gravitational fields (or at least regions thought to contain strong fields) in deep space, where indirect observations can provide clues to the validity of the theory under conditions wildly different from any known on or near Earth. (Of course, we must distinguish between experiments that test the fundamental basis on which Einstein's theory is constructed—for example, the Special Theory of Relativity—and those that test consequences of the theory itself, such as the advance of the perihelion of Mercury.) To discuss all these tests is beyond the scope of the present discussion. (For a valuable review of the experimental tests, see the article written by Clifford Will.)[5] For now, let me note that the classical tests, such as the deflection of starlight, support General Relativity at about the 10 percent level, whereas other tests, such as the perihelion advance of Mercury (and other planets), the gravitational red shift (discussed by Robert Vessot), and the more recent measurements of propagation of radar past the Sun, all support General Relativity at the 1 percent level. However, all of these tests are done in an extremely weak gravitational field, our own solar system.

Progress in testing Einstein's ideas outside the solar system has been relatively more exciting and rewarding, both qualitatively and quantitatively, during the past decade. A few examples will illustrate. Special Relativity is based on two postulates: the principle of relativity, which asserts that physics is the same in reference frames moving at constant velocity with respect to one another, and the proposition that the speed of light is independent of the velocity of its source. This latter statement leads to the predictions at variance with common experience, such as the dependence on motion of the ticking rate of a clock. Until recently this theoretical prediction had received experimental confirmation only at a level of about one-tenth of a percent. However, the discovery of pulsating binary x-ray sources in our own and neighboring galaxies has allowed an unforeseen test of this idea, with extraordinary precision. These sources are seen to emit x-rays with very regular pulse periods of less than a second to a few minutes. The x-rays are emitted from a rapidly rotating neutron star, which itself is orbiting around a companion star. Now imagine that the propagation speed of these x-rays depended on the motion of their source. If the velocity of the x-rays (like light) depended on the speed of the source at the moment of emission, then those pulses emitted when the source moved toward the observer on Earth might get here before those x-rays emitted when the source was moving away from us on the previous orbit. No such peculiar events are seen from these sources. From this analysis, my own studies have found that the speed of light (in this case, x-rays) must indeed be a constant, with an accuracy of better than one part in a billion.[6]

So Special Relativity is well tested and valid beyond the solar system. Other bases of General Relativity, such as the equivalence principle, have been pushed to even higher precision on Earth by Robert Dicke, who has demonstrated the equivalence of inertial and gravitational mass with a precision of better than one part in 100 billion.[7]

An even more spectacular weak-field-gravitation laboratory has been provided for us by nature in the guise of another binary star system, this one containing a radio pulsar called PSR 1913+16. This system appears to consist of two collapsed stars moving about one another in a highly noncircular orbit, with an orbital period of about eight hours. This system has confirmed the precession advance, again with a precision of about

1 percent. Even more important, it has provided the first confirmation of one of the most dramatic predictions of General Relativity. Just as Newtonian gravitation had to be modified because of the finite propagation effects of the speed of light, it was clear from the outset that in General Relativity, gravity should propagate in a wavelike motion similar to electromagnetic radiation. Early work by Einstein and his collaborators had shown that the strength of such gravitational radiation from terrestrial sources would be so weak as to be virtually undetectable directly. Indeed no gravitational radiation, terrestrial or extraterrestrial, has been detected to date. Nonetheless the effects of the existence of such radiation directly manifest themselves in a binary star system by the gradual loss of energy of the system. (Gravitational waves, produced by the motion of two stars, would carry energy away.) In the case of the binary pulsar, this means that the stars gradually move together, decreasing the orbital period, and this effect has been seen, perhaps providing the strongest confirmation of the theory to date.

It is probably fair to say that General Relativity has become the successor to Newtonian gravitation as the best description of space and time. But is it the ultimate theory? Is it applicable everywhere and in every time frame? To this question one must answer, as did Einstein himself, probably not. "The present theory of relativity," wrote Einstein, "is based on a division of physical reality into a metric field [gravitation] on the one hand, and into an electromagnetic field and matter on the other hand. In reality space will probably be of a uniform character and the present theory be valid only as a limiting case. For large densities of field and matter, the field equations and even the field variables which enter into them will have no real significance. One may not, therefore, assume the validity of the equations for very high density of field and matter, and one may not conclude that the beginning of the expansion must mean a singularity in the mathematical sense. All we have to realize is that the equations may not be continued over such regions."

These were the last words that Einstein wrote on the subject before his death in 1955. This is not the discouraged resignation of a man about to face the end of his search for understanding; years earlier he had made very similar remarks to his colleagues. Cornelius Lanczos, a collaborator of Einstein in the 1930s, wrote that "it was therefore quite a shock when he said, 'But why should anybody be interested in getting exact solutions of such an ephemeral set of equations?' I remember very well this word 'ephemeral.' It meant that he did not consider his gravitational equations as the last word."

The basic ideas in Einstein's General Theory of Relativity no doubt are closer to the truth about space and time than any ideas that preceded them or any that have followed up until now. But they too must have their limitations. Perhaps Einstein's ideas about space and time even contain the seeds of their destruction. Just as Newtonian gravity is based on notions of absolute space and time, perhaps the more amorphous notions of space and time in relativity theory will find their successor. In its present formulation, General Relativity allows for the existence of singularities at the centers of black holes and at the beginning of the Universe, places that lie outside of the domain of the theory. Singularities usually point to a breakdown in the theory leading to them. Unfortunately the attitude among physicists and astronomers is to accept blindly these features of General Relativity as one further triumph of the theory and to exploit them as explanations of all puzzles in modern astrophysics. Perhaps if we pay more attention to the spirit rather than the letter of Einstein's work, we may see how to make the next great advance in our understanding of nature.

Figure 4
Albert Einstein. (Photo from AIP Niels Bohr Library)

One thing I have learned in a long life: that all our science, measured against reality, is primitive and childlike—and yet it is the most precious thing we have.

—A. Einstein (frontispiece to *Albert Einstein: Creator and Rebel* by B. Hoffman)

References

1. R. W. Clark. *Einstein: The Life and Times*. New York: World Publishing, 1971.

2. A. Einstein. *The Meaning of Relativity*. Princeton: Princeton University Press, 1955.

3. W. S. Adams. "Relativity Displacement of the Spectral Lines in the Companion of Sirius." *Proceedings of the National Academy of Science* 11 (1925): 382.

4. J. L. Greenstein et al. "Effective Temperature, Radius, and Gravitational Redshift of Sirius B." *Astrophysical Journal* 169 (1971): 563.

5. C. M. Will. In *General Relativity*, S. W. Hawking and W. Israel, eds. Cambridge University Press, 1979.

6. K. Brecher. "Is the Speed of Light Independent of the Velocity of the Source?" *Physical Review Letters* 39 (1977): 1051.

7. R. H. Dicke. *Theoretical Significance of Experimental Relativity*. New York: Gordon & Breach, 1964.

ROBERT F. C. VESSOT

ROCKETS, CLOCKS, AND GRAVITY

Perhaps I should begin by stating that the relationship between experimentalists and theorists is not necessarily an adversary one. Although the theorist and the experimentalist do look at physics in somewhat different ways, the differences lie chiefly in how we work rather than in our understanding of the subject.

Probably in no area of physics is this difference of emphasis more evident than in the study of gravitation and relativity. The mathematical manipulation describing the behavior of space-time when massive bodies are moving with relation to each other is complicated and requires a high level of computational ability. On the other hand, it is not too difficult to understand the general philosophical picture. It is certainly reasonable to accept Einstein's view of the physical universe as an extension of the Newtonian picture that has governed our thinking for over two hundred years. As for any theory, however, such a view is subject to proof—and this is the role of the experimentalist.

In contrast to the theorist, the experimentalist often thinks of physics in terms of technology and hardware. This is definitely not the view of the early natural philosophers schooled in the Aristotelian tradition, where any kind of "hands-on" practical approach was very much beneath their dignity. Perhaps an example of an early experimental philosopher is the person who first took a bucket of water and swung it over his head and wondered why the water did not pour out.

Another view of this difference may be the distinction between the natural philosopher who observes nature and tries to understand what is happening and the experimental philosopher who makes something happen in a controlled way and then measures the relationship between cause and effect, such as swinging a bucket at a known speed over a given radius. We should also recognize that the experimentalist must have some of the characteristics of the artisan in order to provide apparatus and machinery to aid in his measurements and observations. Kenneth Brecher shows that some of the principles behind Einstein's theory are not really so arcane and difficult to accept as we may once have believed. However, it is still very difficult to prove by experiment.

In the seventy years since Einstein announced his General Theory of Relativity, we have become used to his idea of an expanding universe still in a state of evolution. Today our attention is focused on the way this universe is evolving: Is it forever expanding or eventually contracting? How did the primordial eruption of creation occur? How will it all end?

The tests that could confirm what was a radical new theory seven decades ago remain very few and relatively imprecise when compared to those in other branches of physics. In fact only recently have tests been made beyond the 1 percent level of accuracy. Our physical surroundings, even when expanded to include the entire solar system, provide gravitational and relativistic effects that are extremely small. Accurate measurement of these effects in a controlled way, even by use of the immense masses (in our scale of thinking) of the Earth and Sun, only recently has become possible as a result of our evolving technology. The space program has conquered gravitation to the extent that we can now use the entire solar system as a laboratory for controlled experiments.

The magnitude of the effects of Einstein's theory for objects in our solar system is very small. The deflection of light rays grazing the Sun is 1.75 seconds of arc, roughly the angle subtended by a handbreadth over a distance of 11 miles. (An arc second is a measure of angle. There are 3600 arc seconds in 1 degree of angle.) The slowing of time at the surface of the Sun is two parts in a million, or about 1 second in 5.8 days. For the Earth, the effect is much smaller, about 7 parts in 10 billion or 1 second in 45 years. Testing these small effects at a level of accuracy as modest as 1 percent requires apparatus of very high precision and considerable effort.

Microwave and radar technology has made it possible to bounce signals off distant planets and communicate over interplanetary distances. Pulsed laser energy allows us to bounce signals off the Moon from reflectors placed there by astronauts as part of the Apollo program. The technology of clocks and frequency standards is now advanced to levels where we can observe time differences, in measurements made over intervals of about 1 hour, that are equivalent to 1 second in 50 million years. This has changed our view of measurement techniques, and clocks are now the basis of a new type of metrology that allows us to make time measurements with an accuracy requiring up to fifteen places of decimals. The new metrology, based on time and frequency, allows measurement of distance, velocity, and angle, as well as other physical quantities.

The notion of expressing distance in terms of time has long been familiar to astronomers. For them the light-year is a very conventional concept; for most of us it leads to mind-boggling concepts of distance. A light-year is the distance traveled by a light pulse over a year; light moves very fast, approximately 186,000 miles per second (very nearly 300 million meters per second). The velocity of light in a vacuum is a constant of nature and, as Kenneth Brecher discusses in the context of General Relativity, it is a highly fundamental constant that has no dependence on the relative velocity of source to observer. The light rays from the most distant quasar, at a great distance and moving very fast away from us, travel past us at the same speed as the rays from the glow of a recently lighted cigar or the AM radio signals blaring rock music.

With today's technology we can measure distance in terms of time intervals and do it very accurately over long distances. We do this with clocks that generate precise time intervals and provide highly stable frequencies of radio, microwave, and even light signals. These techniques are fundamental to describing a recent test of relativity, and will illustrate how time and distance are related and how this relationship is used to track spacecraft. (Incidentally the same basic techniques are now used in standards laboratories to define distance, and we can expect that the concept of a unified standard of length and time will soon become accepted as international law.)

Using atomic clocks of extremely high stability to measure both time and distance across large regions of space allows us today to measure directly the phenomena of time warping and space warping, that is, alteration of the rate of time flow and of the geometry of space, which are the manifestations of Einstein's relativity (and also are familiar to those who watch "Star Trek" and other futuristic adventures on television).

Figure 1 shows a transmitter sending microwave signals, whose frequency is 1 billion cycles per second. The transmitter is controlled by an atomic clock whose ticking rate depends on the magnetic properties of the nucleus and outer electron of atoms of cesium 133. It is a device that generates a signal at 9,192,631,770 cycles per second. In other words, the device

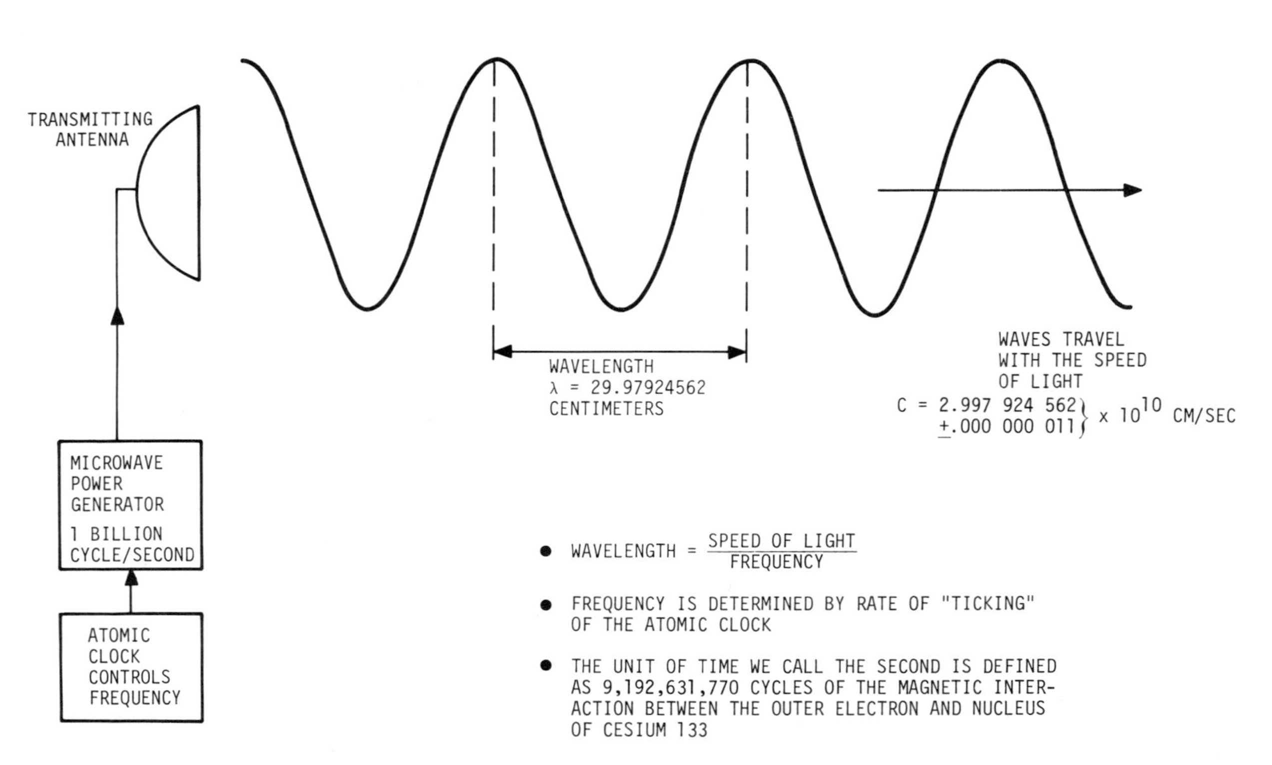

Figure 1
How time intervals define distance.

depends on a specific type of atom, all of which are alike in nature, and therefore is universally accessible as a fundamental standard. (Currently the internationally accepted definition of a second of time is 9,192,631,770 cycles in the magnetic interaction between the outer electron and nucleus of cesium 133.) Other atoms are suitable for clocks, for instance, hydrogen, which is the basis of the atomic hydrogen maser clocks with which I have been working. The principle of operation is similar, but the stability is substantially higher.

In the case of the clock in figure 1, we are generating 1 billion cycles per second, and our transmitter is broadcasting radio waves at the same 1 billion cycles per second. The waves travel from the antenna with the velocity of light. Because of the constancy of the velocity of light, the distance from crest to crest (the wavelength) depends only on the rate at which the waves are generated.

To use these properties of time stability and constancy of propagation of signals to measure distance requires that we receive and count these waves as they arrive; each complete wave received counts as one cycle. Figure 2 shows an arrangement to count waves as they are received. In our example, the billion cycles per second received are counted by comparing their numbers to those of another clock, which is also generating a billion per second. If the receiver is moved toward the transmitter, we find that an extra cycle is gained for each wavelength of distance moved. This is the result of shortening the distance, which contains a fixed number of cycles. If we move away from the transmitter, we must allow for the inclusion of an additional cycle for each wavelength of distance we move, and hence we will lose the cycle at the receiver. Since the number of cycles generated is exactly known, we can measure changes of distance very accurately over very long distances simply by counting the gain or loss of cycles using another atomic clock at the receiver. The limitation on this technique lies only in the ability to detect the signals, for today's powerful transmitters and large antennas enable us to send and receive signals anywhere in the solar system and beyond.

This one-way distance-measuring system requires two clocks. In the last few years, however, atomic clocks have become small, rugged, and reliable enough to put into spacecraft. Thus the conventional tracking system today is a two-way system (shown in figure 3), which requires only one clock. Here we send signals

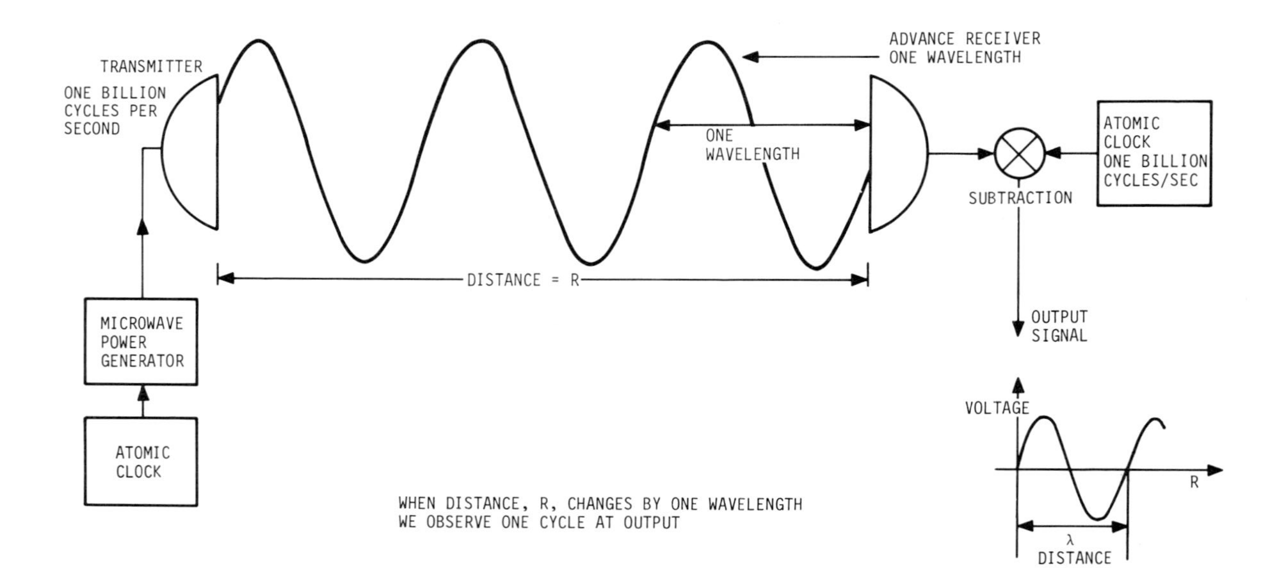

Figure 2
Concept of one-way signal paths for measuring distance.

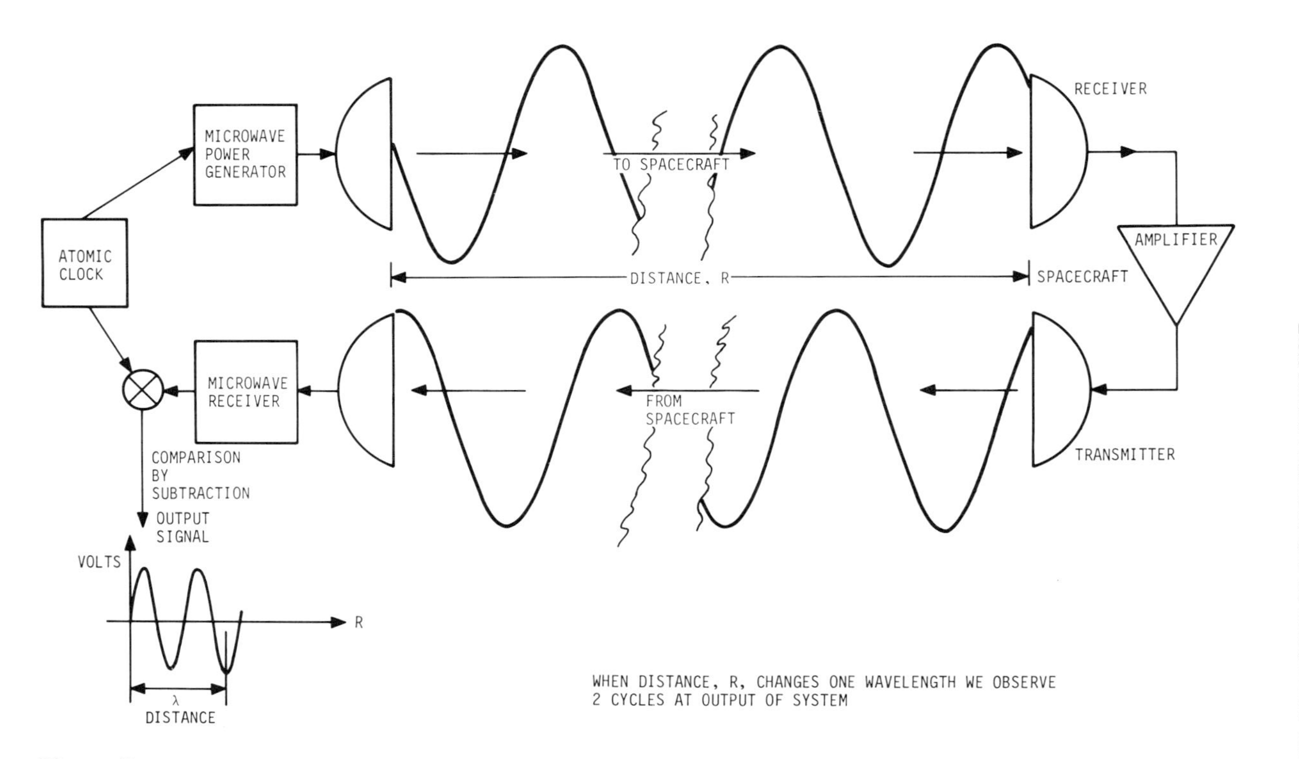

Figure 3
Concept of two-way signal paths for measuring distance.

to the space vehicle and retransmit them back to the Earth station and use the same clock both to control the transmitter and to count the received cycles. Because waves will go to and come from the spacecraft, every time the distance changes by one wavelength, we detect two cycles at the output. Over short distances it is sufficient simply to reflect the signal. This is the basis for Doppler radars familiar to anyone caught speeding on the highways. For longer distances, say, at the distance of the Moon, the return signal must be retransmitted, or boosted, by an on-board transceiver.

The accuracy of these systems, if limited only by clocks, is extraordinarily good. If we average our measurements for about fifteen minutes, we can see changes of distances as small as seven-thousandths of an inch. The significance for testing theories of relativity is that this same measurement can be made over tremendous distances.

We can also use clocks to measure astronomical angles as well as distances by a technique known as Very Long Baseline Interferometry. In this technique, two receivers are spaced many thousands of miles apart, with each listening to the signal from a common source in the sky. These cosmic signals (most often radio waves) may be thought of as long strings of waves, with a different string connecting each receiver to the source. If the receivers are not equidistant from the source, the strings will include a different number of total wavelengths. Determining this difference by combining the recordings from each station, we can measure the angle between the baseline (the line separating the two receivers) and the direction to the source. (For example, a 90° angle means the source is equidistant from the two receivers, i.e., the pulse arrival times are the same.) The sensitivity in the angle measurement currently available is about 10^{-4} seconds of arc, which is about the angle subtended by a dollar bill seen on the surface of the Moon. (The technology is improving at a fast enough rate that we can almost keep track of the shrinking dollar!)

In the last two decades this technique of angle measurement, wherein we listen to stars, quasars, and galaxies that emit radio noise, has opened a new field of astronomy. It has also provided some precise tests of relativity; in particular, it has allowed us to see how radio waves are bent when they go by the Sun. Each October, two quasars, 3C273 and 3C279, both powerful radio noise sources, are seen near the direction of the Sun. One of them actually goes behind the Sun and the other is reasonably close, so that the angle between them can be measured as their line-of-sight approaches the Sun. This Oktoberfest of data has confirmed the bending predicted by Relativity Theory at a level of approximately one part in a thousand and the study is being continued by Richard Sramek of the National Radio Astronomy Observatory.

Experiments of this type also have been done by looking at radio transmitters both placed on Mars and orbiting that planet as part of NASA's Mariner and Viking programs. Here both the deflection and the time delay of radio pulses can be measured as the rays pass close to the Sun, and Irwin Shapiro of the Massachusetts Institute of Technology has data at better than one part in a thousand, again confirming the theory.

An experiment of particular interest to me is the measurement of the gravitational effect on time itself. On June 18, 1976, in a joint program involving NASA and the Smithsonian Institution, we launched a rocket containing a very stable clock straight up from Earth to a distance of 10,000 kilometers. The object of the experiment was to see how time is affected by gravity as the clock went away from, and later fell toward, the Earth. According to Einstein, the ticking of a clock, when it is at its farthest distance from the Earth, should be some 4 parts in 10 billion faster than when it is at the Earth's surface.

In order to see the ticks of the spaceborne clock, we used microwave signals (instead of the light pulses Einstein prescribed some seventy years ago). The process of counting the relative rate of the clocks would have been easy if the spacecraft had been kept at a constant distance from Earth. Our situation was different because we used a space vehicle that was shot upward and was in free-fall over most of its trajectory. The path distance between the Earth station and the vehicle was constantly changing and, as shown in figure 2, there is the complication of accounting for all the cycles or wavelengths of the microwave signals that lie in the path. To overcome this problem, we used the system shown in figure 3, which allowed us to count the wavelengths in the two-way (up-down) signal path using the ground clock. By dividing the number of these wavelengths by two, we determined the number of one-way wavelengths at any given instant. We then subtracted this number of cycles from that emitted by a clock mounted on the spacecraft, using the one-way system of figure 2. The difference in cycle counts measures the difference in clock-ticking rates due to the effects of General Relativity. In fact, with this technique we can identify two relativistic effects: one the result of the relative velocity of the clocks, the other the result of having a difference in the strength of gravity at the locations of the two clocks. The latter effect is called the gravitational red shift and is predicted by Einstein's General Theory; the former is predicted by Einstein's Special Theory.

The hardware used in the experiment is important. Remember that technology depends on artisans—mechanics, pipefitters, electricians, and clock builders—to do this kind of work. Figure 4 shows Martin Levine, my coinvestigator, and me holding our then new, lightweight (90-pound) hydrogen maser clock for space use. This clock was placed in a space-probe vehicle

Figure 4
M. W. Levine (left) and R. F. C. Vessot (right) holding the hydrogen maser clock developed for their 1976 space-probe test of relativity. (Photograph by Charles Hanson, Smithsonian Astrophysical Observatory)

(figure 5) and attached at the top of a four-stage solid-fuel Scout rocket. The assembled rocket was then launched.

The vehicle was tracked very accurately so that we could predict the relativistic effects to levels consistent with the stability of the clocks. This required knowing the probe's location to 100 meters and its velocity to within 2.5 centimeters per second at all points along its path. The track is shown in figure 6. The retransmitted signal from the space probe was also received at three other locations besides the one shown at MILA (Merritt Island, Florida), where the ground-based hydrogen maser clocks were located. These other stations were Bermuda, Ascension Island, and Greenbelt, Maryland; the last is shown as NTTF (National Test and Training Facility, Goddard Space Center). The data from all four stations were used to calculate the probe's trajectory.

Some of the experiment results are shown in figure 7, which consists of sections of strip charts moving at two major divisions per second. Both the laboratory and the spaceborne clocks may be considered as emitting an oscillating signal. The difference between the two oscillating signals is plotted in figure 7. A flat curve means the two clocks are ticking at almost equal rates. The top curve in figure 7 is taken from a time in the flight when the rocket had slowed to a point where the amount of time change due to the relative velocity of Earth and rocket is almost equal and opposite to that due to the gravitational red shift. As a result, the curve is quite stretched out. As the probe continued to ascend and reduce its speed, it reached the maximum altitude of approximately 10,000 kilometers. The bottom curve in figure 7 refers to this moment in time. Now the rocket is at its slowest point, and almost all the difference in clock rates is due to the gravitational red shift. We see the full extent of the speeding up of the spaceborne clock in its weaker gravitational field. This is just what we expected; the clock's ticking rate has increased by four parts in 10 billion. The process reverses as the space probe falls. Finally the received signal is broken up as the space probe goes below the horizon of the main tracking station and, of course, ceases when the probe plunges into the ocean.

When we compared these results with Einstein's predictions, we found agreement at the seventy parts per million level. That is, our proof agreed with prediction within 0.000070 of the magnitude of the total observed effect. While the effect was very small, the clocks and microwave system permitted a com-

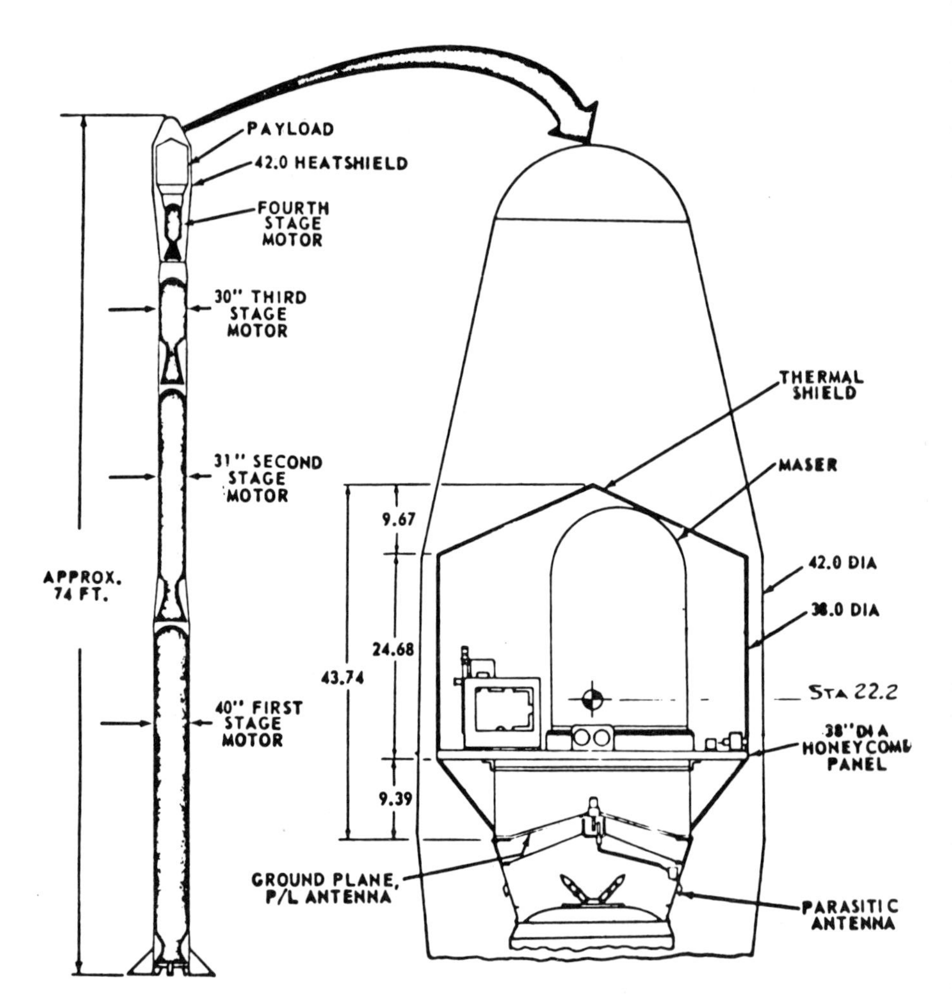

Figure 5
Scout vehicle for space-probe test of relativity.

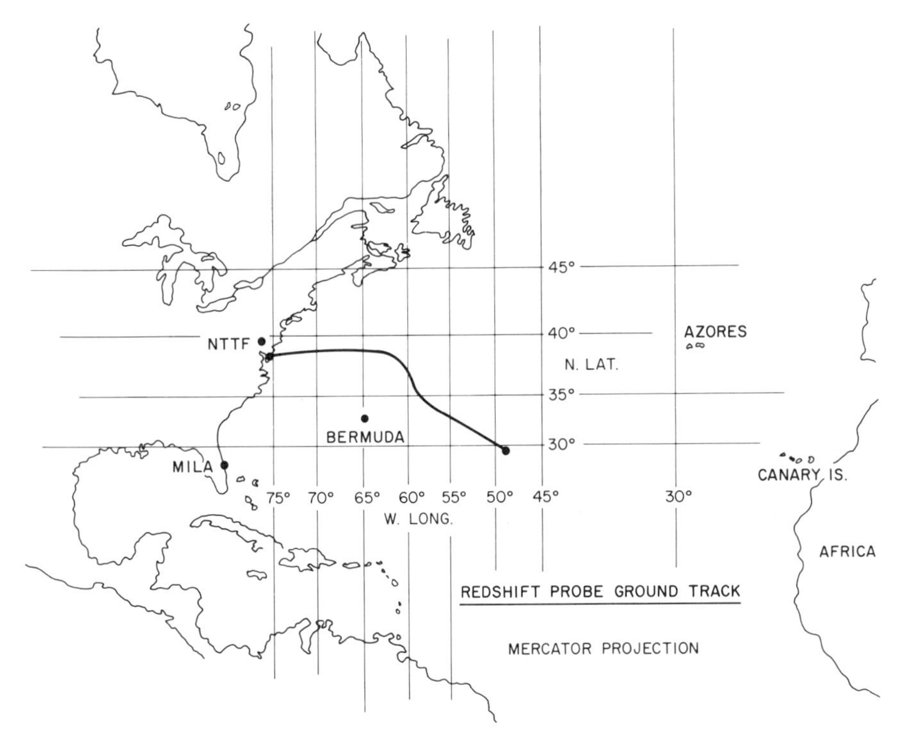

Figure 6
Projection of trajectory of payload on the Earth's surface.

parison of time over the farthest part of the mission at a precision of seven parts in a million-billion (7×10^{-15}), giving us the resolution we required.

Our test has given us considerable confidence in Einstein's predictions. It improved by a factor of more than one hundred on the results of an elegant and beautiful experiment made at Harvard's Jefferson Laboratory by Professor Robert Pound in 1956. His tests used a precisely defined wavelength of emitted gamma rays to measure the gravitational red shift between source and detector over a vertical separation of about 25 meters. The technology in each of these experiments was totally different, yet each agreed with Einstein within its limits of detection.

Where do we go from here? The accuracy of clock tests on Earth may be pushed further, perhaps to one part per million, but the Earth's mass is so small we must look for bigger masses, like that of the Sun. The expected effect of the mass of the Sun,

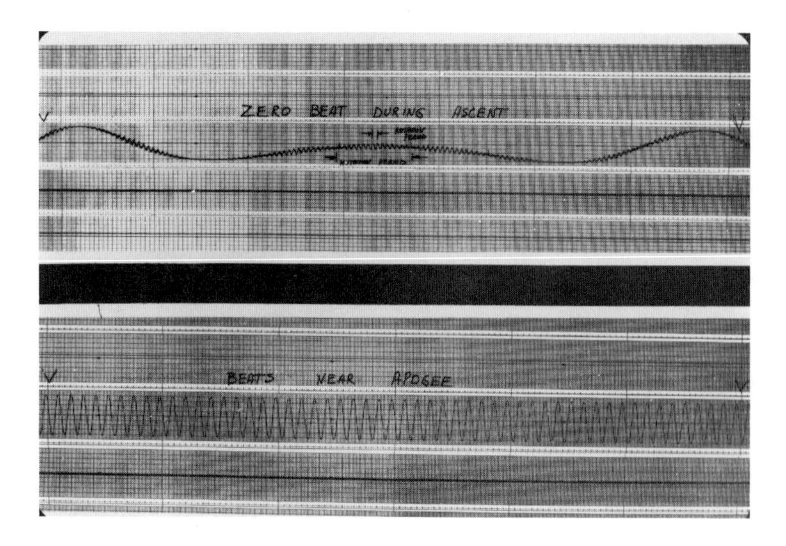

Figure 7
During ascent (upper graph) a time exists where the apparent slowing of the space clock owing to the rocket velocity is compensated by the speeding up of the space clock owing to the decreasing gravitational potential between space and Earth. At apogee (lower graph), where the space vehicle's speed is minimum, the rate of the space clock appears to be about 0.9 tick per second faster than the Earth clock. (Photograph from Smithsonian Astrophysical Observatory)

even at a distance from its surface of three solar radii, is about five parts in 10 million or one second in twenty-three days, much larger than the one second in forty-five years we have from the Earth.

In the past two years there have been serious discussions at NASA about a mission to the Sun, coming within a distance of four solar radii of its surface. The details of that mission, and whether it will carry a clock, are still being discussed. If it could be used as a gravity test, however, much could be learned from such a solar mission. One version of this mission would start from Earth and reach Jupiter in one and a half years. Then using Jupiter's gravity, it would swing around that planet, change direction, and start on a two-year trip to the Sun. The space probe would then fly over the north pole of the Sun and travel from north to south in less than fourteen hours. The test of the time warping (or gravitational red shift) could be done at a level of one part in 100 million. We could then look for the actual shape of the warping sufficiently accurately to distinguish between Einstein's theory and some newer theories that are similar to his in their coarser details. And once we have accurately tested the theory, we can use it to learn about the interior of the Sun.

In fact, there are few other ways to obtain information regarding the Sun's interior mass distribution and motion than observing the gravitational effects. The behavior of a free-falling particle, such as a spacecraft equipped to compensate for nongravitational forces of light pressure and particle flux near the Sun, can provide this information if it can be tracked with the required precision. At present, a prime gravitational objective of the solar mission, tentatively called STARPROBE, would be to measure the solar oblateness, that is, the flattening of the Sun caused by the centrifugal effects of its rotation. This effect is not measurable from the optically perceived roundness of the solar disk and, because of the complicated movements of the solar matter, actually may cause a more complicated distribution of mass than that due to simple rotation. In addition to the mass distribution, it is theoretically possible to determine direct effects from the rotation of the Sun. General Relativity predicts that the effect of a spinning massive body is to twist the geometry of space around the body, over and above the stretching caused by the gravitational red shift of a nonrotating massive body. Since the usefulness of these measurements to the science of solar and stellar physics depends on gravity theory, our first objective must be to test this theory to sufficient accuracy to validate our data.

This process of using measurements based on a well-tested theory in one area of physics to obtain information for use in another area is typical of the experimental method where observation leads to theory. Theory leads to prediction and prediction, in turn, demands proof. Having proof of a theory, we can use it as a basis for further observation and thus continue the cycle.

Studies of the Sun are not merely of academic interest. For humankind, the Sun is a vital part of existence. Any slight variation in its behavior could lead to dramatic climatic change—another Ice Age perhaps, or, conversely, a period of intense and destructive heating. Some foreknowledge of the Sun's behavior, especially if it has any prediction for change, could allow us intelligently to alter our ways of life and general system of values to survive such changes. The ongoing process of evolution, in which we are now participating, will at some time require for our survival an extensive understanding of the Sun's behavior. The forthcoming STARPROBE program, using a test of theory, is the beginning.

Further Reading

Calder, Nigel. *Einstein's Universe.* New York: Viking Press, 1979.

French, A. P., ed., *Einstein: A Centenary Volume.* Cambridge, Mass.: Harvard University Press, 1979.

Einstein, Albert. *Ideas and Opinions.* New York: Crown, 1954.

Hoffman, Banesh, and Dukas, Helen. *Albert Einstein: Creator and Rebel.* New York: Viking, 1972.

Rindler, Wolfgang. *Essential Relativity,* 2d ed., New York: Springer-Verlag, 1980.

CHAPTER 3

THE EVOLUTION OF THE SOLAR SYSTEM

FRED L. WHIPPLE

THE MATERIAL INGREDIENTS AND STRUCTURE

Observation and theory are so intricately entwined when related to the origin of the solar system that I will make no attempt to untie the Gordian knot. Moreover, it would be absurd to present tables of the planetary observations that led to statements about the orbits and masses of the planets or other such data. Such observations, according to older theories, proved that the Earth was the center of the universe. Rather I shall present only those observational data, or fundamental facts, about the solar system based on theories adequately supported by experimental or observational data.

Among major bodies of the solar system, the Sun, the planets, the satellites, and the miniplanets known as asteroids, we find a highly flattened system with revolution in a common direction (figures 1 and 2). Except for Pluto and Neptune, the planet orbits are well separated, with the spaces increasing with increasing solar distance. This is known as Bode's Law. The Sun and six of the nine planets also rotate in the same sense. The several exceptions to this rule allow for a range of solutions and, for astronomers, add to the interest of the problem. Only the comets (figure 3) move in random directions with random rotations as they come in from the Oort cloud, a region of the outer solar system extending to perhaps fifty thousand times the Sun's distance from Earth.

This spinning or rotating property of the system, however,

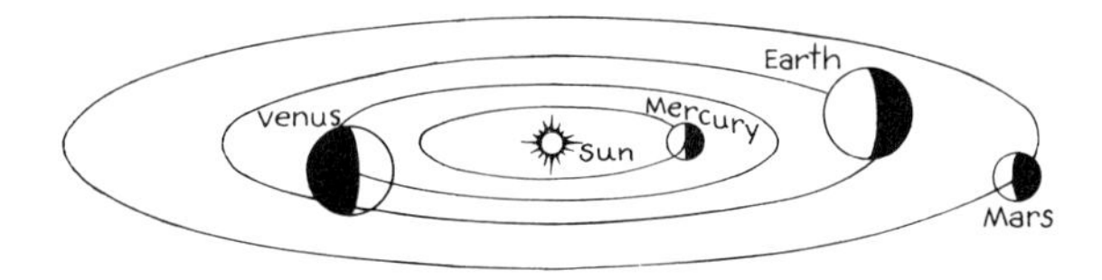

Figure 1
Orbits of the terrestrial planets about the Sun, a projection with planetary diameters grossly exaggerated. (From *Orbiting the Sun,* courtesy Harvard University Press)

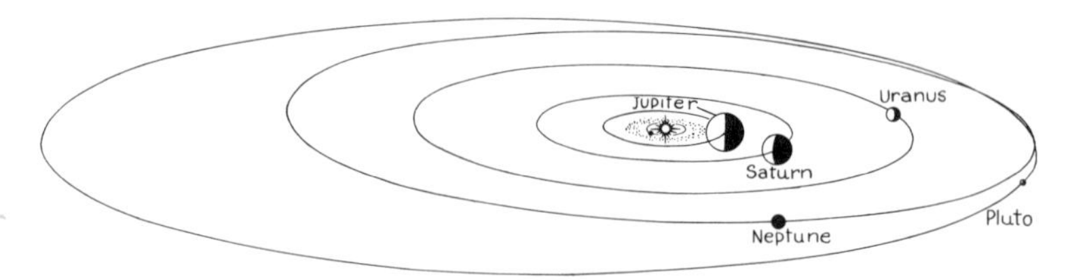

Figure 2
Orbits of the outer planets. The innermost orbit is that of Mars (see figure 1), with the asteroids lying between it and Jupiter. (From *Orbiting the Sun,* courtesy Harvard University Press)

Figure 3
Comet West (1976) with its irregular ion tail (above) and smooth dust tails (below). (Courtesy Jack W. Harvey, Kitt Peak National Observatory)

has a most improbable characteristic. Whereas the Sun contains almost 99 percent of the mass, Jupiter, with only a thousandth as much, carries nearly 60 percent of the rotation, or angular momentum, of the entire system. For a body in a circular orbit about a central mass such as the Sun, the angular momentum is a product of the mass, distance, and velocity of motion, and is constant during the orbit, as Kepler noted. The four giant planets together contribute about 99 percent of this rotation or angular momentum, whereas the Sun, through the spinning about its axis, contributes only about one-half of a percent. Were the angular momentum to be distributed in what we might consider to be a rational fashion, in proportion to the mass, the Sun would spin much faster, with a period of only a few hours, instead of more than three weeks. Explaining this unlikely situation is one of the major challenges for theories of the origin of our solar system.

In the last two decades the magnificent space program of planetary exploration and the advances in observational astrophysics have given us an entirely new insight as to the nature of the planets and the manner in which stars and planetary systems originate and develop. We now know more about the interior of the Moon, Mars, and even Jupiter than we knew about the Earth at the turn of this century. Furthermore studies show that star formation in our galaxy occurs by the gravitational collapse of huge masses and volumes of interstellar gas and dust. Figure 4 shows such a system, Eta Carinae, where new stars are being formed. There are many such stellar incubators busily hatching stars in many parts of our galaxy and in other similar galaxies throughout the universe. At the same time, we find no evidence for stars being formed singly, in isolated regions.

With our newly gained knowledge of the composition of the planets and of the basic nature of stellar origin through a collapse of a great interstellar cloud, we are in the position of having analyzed the ingredients of a great meal prepared by a cosmic cook. Having studied the finished product, we know the number and proportion of the ingredients distributed among the various bodies and how they are layered within these bodies. The grand recipe for making planets and satellites, however, involves more subtle nuances, such as baking, boiling, stirring, separating, decanting, coagulating, peppering, and cooling, as well as establishing time scales for all these operations.

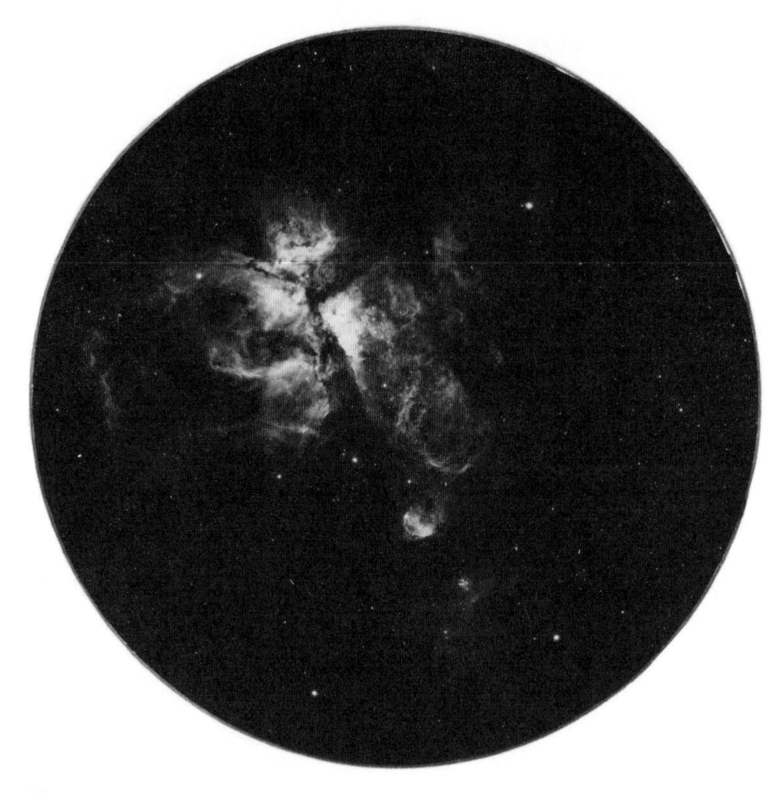

Figure 4
The huge gas and dust cloud of Eta Carinae, illuminated by hot new stars. (Photograph from Harvard College Observatory)

Laboratory studies of radioactive atoms and their decay products in samples of the Earth, meteorites, and the Moon give us one fact of extreme importance, about which there is no remaining doubt: the time scale of the cosmic cooking process. Table 1 shows a few of the most important atomic clocks used in determining the age of the solar system.

Consider the radioactive atoms of uranium and thorium, which decay into leads of different atomic weights plus helium. Or consider potassium 40, which leaves a daughter product of argon 40. Since these daughter gases continue to escape from any molten material, the atomic clock is set when the material freezes. If we measure the amount of potassium 40 and argon 40 remaining in a lunar or meteoritical sample obtained today, we can calculate the number of years since that sample froze. In other words, we can determine the age of the solid body. For

Table 1
Selected radioactive isotopes

Atom	Decay Products	Half-life (Millions of Years)
Uranium 238	Lead 206 + He^4	4,510
Uranium 235	Lead 207 + He^4	704
Thorium 232	Lead 208 + He^4	14,000
Iodine 129	Xenon 129	17
Rubidium 87	Strontium 87	49,000
Potassium 40	Argon 40 (11%)	1,300
Aluminum 26	Magnesium 26	0.74
Plutonium 244	Xenon 131–136	82

large, enclosed bodies where there may have been many chemical reactions and various melting processes, the comparison among the various lead isotopes and among the rubidium-strontium isotopes can give us the age when the masses of material were assembled. Until the late 1950s the age of the Earth, as calculated theoretically, doubled every fifteen years. (See figure 8, chapter 7.) Fortunately this rapid aging was finally halted by the laboratory analyses of meteorites and Earth materials. The Earth and meteorites were first formed about 4.6 aeons (billions of years) ago. I think we were all greatly pleased and somewhat relieved when the lunar rocks returned by the Apollo astronauts and the Russian missions gave the same age for the Moon. Hence we now know the birthdate of planetary formation, 4.6 aeons ago, with an uncertainty of perhaps a tenth of an aeon, or about 100 million years. The Sun was born at roughly the same time, but its age determination is not accurate enough to tell us whether its origin coincided with that of the planets. (Incidentally, the calculated age of the universe has doubled again since 1964 but seems to have reached maturity at somewhat over 10 aeons.)

Heavy hydrogen or deuterium tells another very important tale about the origin of the planets. In the Earth, meteorites, and Moon, the deuterium atoms number about five in a million compared to ordinary hydrogen. In the Sun, the ratio is reduced by a factor of sixty times. Deuterium is a very stable element so long as it is not heated to millions of degrees, whereupon it disintegrates. We should be thankful for this property because deuterium is our major hope for a fusion energy source. But the Sun and stars are hot enough to destroy deuterium by thermonuclear reactions at relatively shallow depths below their surfaces. If the material that made the Earth or meteorites were scooped out of a star, it could not have such a high deuterium-hydrogen ratio. Lithium atoms tell the same story. We can conclude confidently that the planets were not formed according to any of the hypotheses involving tide raising on the Sun, collisions of stars, or other theories that would bring the material out of stars and condense it into planets around our Sun.

On the other hand, the rare form of carbon with an atomic weight of thirteen units (the dominant species, or isotope, of carbon has an atomic weight of fourteen units) makes up about one-ninetieth of all carbon on the Sun, Earth, Mars, Jupiter, meteorites, and the comets. In the case of the carbon isotopes, the ratio is not much affected by mild thermonuclear reactions in the Sun or similar stars. It varies widely, however, in interstellar space. The carbon isotope ratio and, much more importantly, many comparisons among the abundances of the chemical elements show that the Sun and the solar system were all made from about the same primary cosmic ingredients. This fact encourages us to entertain theories in which the Sun and planets were formed from the same source of materials, perhaps simultaneously. The ingredients used by the cosmic cook have now been discovered.

The mix of these ingredients, or elements, in the Sun, which is typical of many stars, has a remarkable property, first pointed out many years ago by Harrison Brown: the elements can be divided into three classes as measured by their freezing points. Table 2 shows this distribution. If our solar system starts out hot enough, everything is gaseous. Nothing much freezes out until the system gets down to temperatures of the order of 2000 K, or some 2000° above absolute zero. Most of the metals and heavy elements, or their compounds, become solids by about 1000 K, making rocks or earthy material. As the mix cools, the next big step occurs when ordinary water freezes at about 273 K (corresponding to 0°C). At lower temperatures, we freeze out ammonia (NH_3), carbon dioxide (CO_2), and finally, at extremely low temperatures with moderate pressure, perhaps even methane (CH_4). By 10 to 50 K, or −263°C to −223°C, we find that we have frozen out everything but hydrogen, helium, and some of the noble gases. (There may be no temperature

Table 2
The solar mix of elements

	Percent by Mass		Factor to Give Source Mass
Gases			
Hydrogen	76.8		
Helium	21.2	98.2	1
Others	0.2		
Ices			
Carbon + 4H	0.45		
Nitrogen + 3H	0.15	1.38	—
Oxygen + 2H[a]	0.77		
Rocks			
Magnesium + O	0.103		
Silicon + O	0.106		
Sulphur + O	0.058	0.45	220
Iron + O	0.143		
Others	0.039		
Ices + rocks = cometary		1.83	55

Source: Derived from A. G. W. Cameron, *Space Science Review* 515 (1973):121.
[a]Oxygen as listed is depleted by one atom for each rocky atom.

anywhere in the galaxy low enough to freeze out the highly volatile hydrogen and helium.)

The last column in table 2 gives the reciprocal of the fraction of the solar mix in each class of elements. For example, about one-half of 1 percent of the solar mix consists of rocky material, of which the Earth is composed. Thus we would need originally some 220 Earth masses of solar mix from which to freeze out one Earth mass. For a comet, made of ices plus rocks, the percentage is less than 2, so the starting mass must add up to some fifty-five times the mass of the comet. Let us now see how this peculiar freezing property of the elements is reflected in the composition of the planets.

The masses and dimensions of the planets give us their mean density. For the Earth and Moon, we learn about the internal structure directly from earthquakes and moonquakes (figure 5). By analogy, and from the character of the surfaces, we can learn

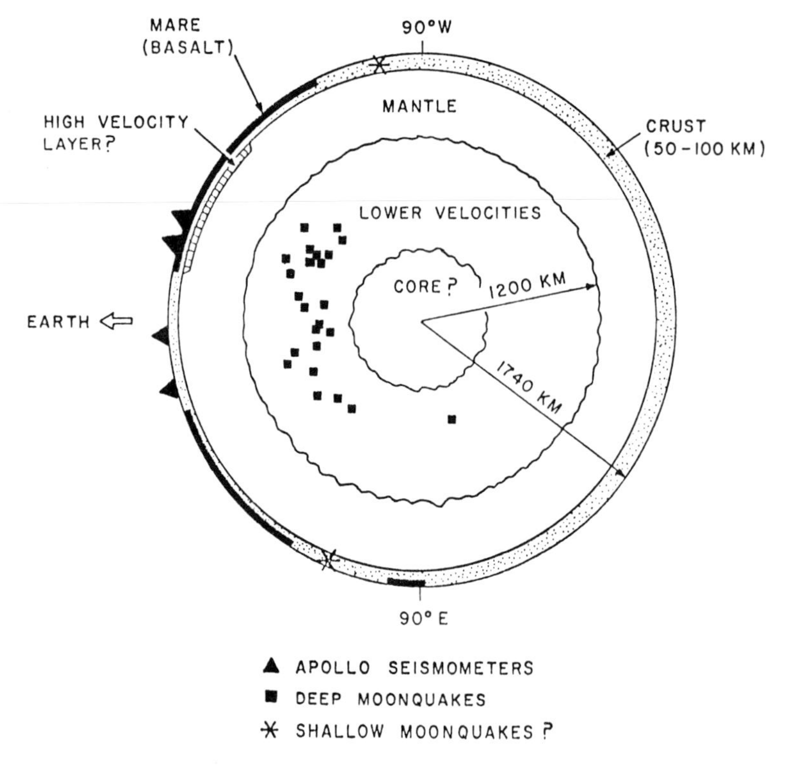

Figure 5
Cross-section of the Moon. The depths of the repeating moonquakes are indicated by squares as located by the Apollo seismometers. The quakes are so weak that they can be detected only on the near half of the moon. (From *Orbiting the Sun,* courtesy Harvard University Press)

quite a bit about Mars and Venus. The giant planets bulge conspicuously at their equators because of their rapid rotation. The bulges, in turn, affect the motions of planetary satellites, enabling us to learn about the densities of the upper layers of these planets from gravitational effects. Space probes in orbits about or passing near any planet provide more of this vital information. Finally the theory of gases at extremely high pressures and/or temperatures is very much better understood than the effect of compression on rocks.

Knowing the density of a planet, we can infer an approximate model for its internal composition—for example, denser planets must contain more iron, and so on. Figure 6 shows that Mercury, with a density of over five and a half times water, consists of at least 60 percent iron and nothing but heavy rocky material. The Earth and almost certainly Venus have molten cores made

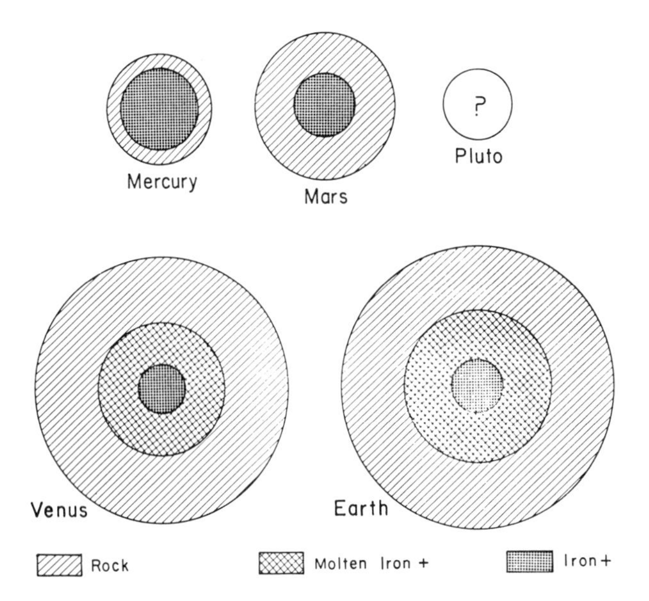

Figure 6
Interiors of the terrestrial planets. The model of Mercury is from Norman Ness; Venus is assumed to be like the Earth; Mars averages several models. (From *Orbiting the Sun,* courtesy Harvard University Press)

up mostly of iron. On Mars a molten core may be smaller than shown in the diagram, but one probably exists. Pluto does not belong among these planets, at least as far as composition is concerned. In fact, the orbit of its satellite Charon suggests a rather low density for the planet, as though it were made entirely of ices and rocks, that is, of cometary stuff.

The terrestrial planets and the asteroids formed inside the orbit of Jupiter are entirely rocky. This is a vital piece of information. The temperature between the proto-sun and Jupiter must have been high enough during the accumulation of the inner planets that water did not freeze. Otherwise water would have been added in great quantities instead of being lightly mixed with or sprinkled on the terrestrial planets, perhaps by comets. The chemistry of the minerals in meteorites shows that they were mostly assembled at temperatures roughly in the range 400 to 550 K, at pressures of perhaps one ten-thousandth of an atmosphere, sometime after the rocky dust had condensed out of the solar mix.

Regarding the large planets (figure 7) a shockingly different situation is apparent. Jupiter is almost a pure solar mix, con-

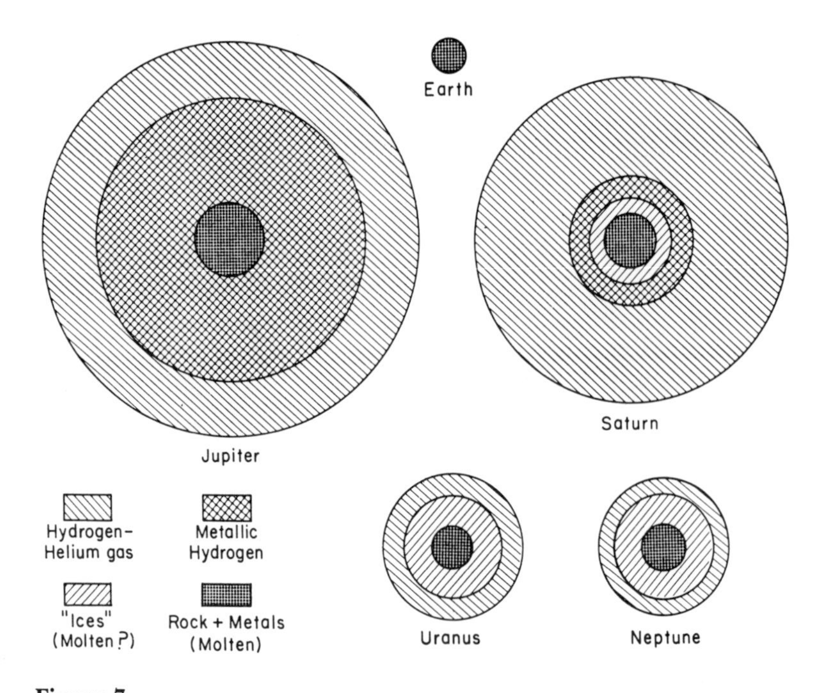

Figure 7
Interiors of the giant planets. The extremely high pressures near the center of Jupiter make the physical modeling somewhat speculative. Jupiter and Saturn models from M. Podolak and A. G. W. Cameron; Uranus and Neptune models from W. B. Hubbard and J. J. MacFarlane. (From *Orbiting the Sun,* courtesy Harvard University Press)

sisting almost totally of hydrogen and helium. Saturn, with less than a third the mass of Jupiter, follows it fairly closely. Uranus and Neptune, on the other hand, have almost exactly the composition we would expect if they were made up of comets. They are the ice-plus-rock mix, with perhaps a little hydrogen and helium added. Returning to Jupiter and Saturn, we find they are composed of the solar mix with an addition of perhaps ten to twenty Earth masses of ices and rocks, about the same amount as the total composition of Uranus and Neptune. The freezing characteristics of the elements tell us that beyond Saturn the building blocks were ices plus rocks—in other words, comets. It is not surprising, then, that we find the comets distributed in a cloud at extremely great distances from the Sun. In the deep freeze of space, the ices could remain unchanged over the aeons since the solar system was formed. How the comets got into these huge orbits with periods of millions of years remains a challenge to theory. Were they formed there, or were they thrown out by perturbations from the giant planets? These questions are still unanswered.

The extraordinary photographs made by NASA's Voyager have provided some remarkable information about Jupiter. The discovery of sulfur volcanoes on Io (figure 8) are among the highlights of these achievements in planetary exploration. Europa and Ganymede, the second and third Galilean satellites, so disturb the motion of Io that its orbit remains somewhat eccentric. Near Jupiter, the accordion effect of gravitationally induced tides heats the interior of Io significantly. This continuous heating over the aeons has boiled away any water Io may once have

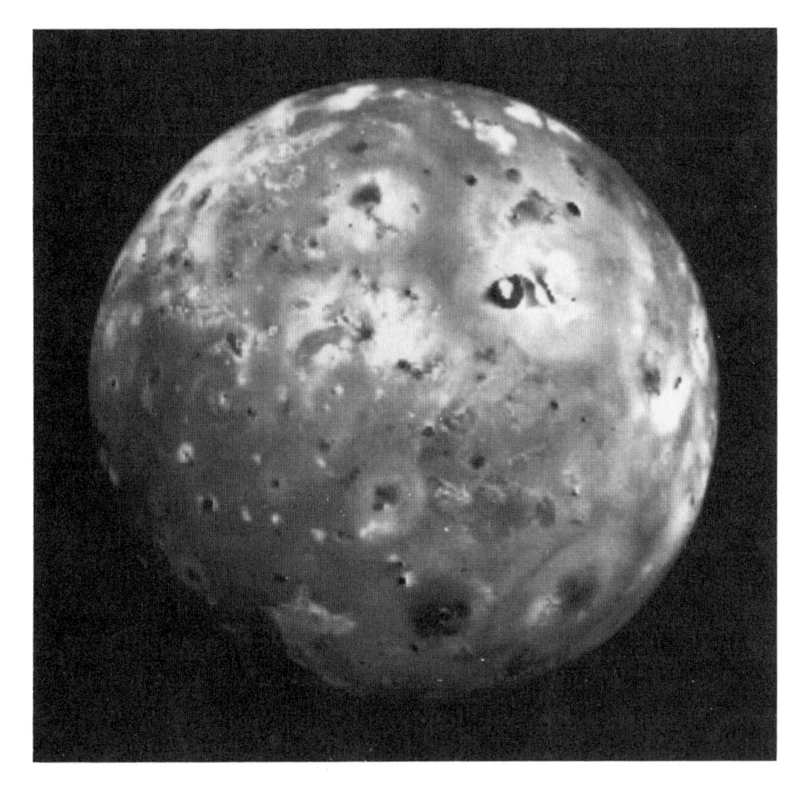

Figure 8
Jupiter's satellite Io as viewed by Voyager 1 in 1975. The formations are various shades of red, yellow, and ochre, the color being derived mostly by sulfur compounds. (Photograph from National Aeronautics and Space Administration)

had and has provided energy to maintain the sulfur volcanoes, of which seven were observed by Voyager 1. Thus Io must be much like the Earth in composition (figure 9), having a molten center maintained by Jupiter tides. Europa is similar, except that the tides are much smaller, one-ninth those of Io, enabling Europa to maintain an icy surface mantle, the smoothest surface in the solar system. In their larger orbits, Ganymede and Callisto contain a much greater fraction of water or ice. Titan of Saturn has even a larger fraction. The constitution of Neptune's Triton remains uncertain.

We can now say something about the minimum amount of matter needed to make the planets out of a solar mix of elements. For the Sun, planets, comets, and asteroids, table 3 lists present masses in terms of the Earth's mass and also the original minimum mass of solar mix from which they could have formed. Nearly ten times the present mass of the planets was needed, amounting, however, to only 1.3 percent of the present Sun's mass. In addition, the Sun may have lost 10 percent of its mass by an early solar gale. Thus the early solar system initially may have contained about 112 percent of the present Sun's mass as an absolute minimum.

Table 3
Masses of solar system bodies (in Earth masses)

	Today	Minimum Original	Assumption re Original
Sun	333,000	370,000	—
Mercury	0.055	23	0.60 Iron
Venus	0.815	180	Rocky
Earth	1.000	220	Rocky
Mars	0.108	24	Rocky
Asteroids	0.001?	10?	Rocky
Jupiter	317.9	1,170	0.05 Cometary
Saturn	95.1	860	0.15 Cometary
Uranus	14.6	800	Cometary
Neptune	17.2	940	Cometary
Pluto	0.003?	<1?	Cometary
Comets	1 + ?	1?	Cometary
All	333,448	374,229	
Planets	448+	4,229+	

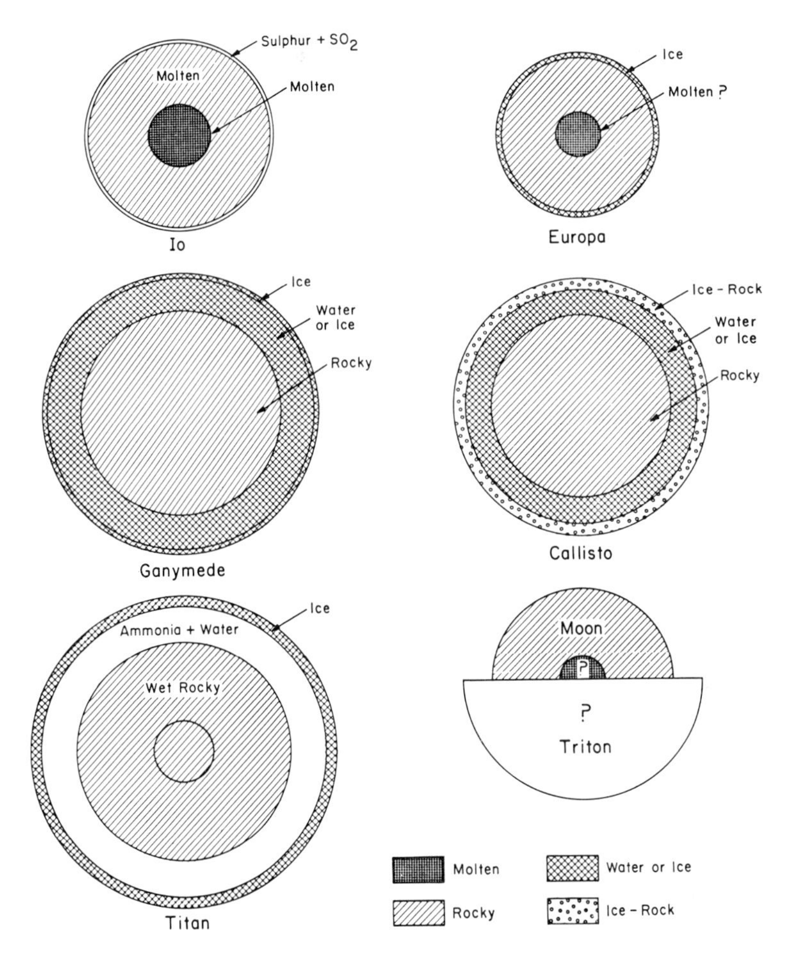

Figure 9
Interiors of Jupiter's Io, Europa, Ganymede, and Callisto and Saturn's Titan, compared to that of the Moon. The Galilean satellites are modeled after Torrence Johnson; Titan after John S. Lewis and Donald M. Hunten. (From *Orbiting the Sun,* courtesy Harvard University Press)

Some minor deviations in the composition of the meteorites, as compared to the Earth and Moon, are extremely exciting. Three lines of evidence point to a remarkable conclusion: that the mix of elements producing the solar system was not a pure sample of the interstellar medium. It was laced with some unusual elements made in a nearby supernova that contributed some of the material rather erratically to parts of the developing solar system. (A supernova is an exploding star, produced when a star undergoes gravitational collapse at the end of its lifetime. During this explosive event the star briefly brightens to many billions of times its normal luminosity and heavy elements are fused from lighter ones in its interior, then blown out into space.) Most of this evidence comes from an unusual and fortunately very large meteorite, the Allende fall in Mexico, 1969. This meteorite was a carbonaceous chondrite, a rare and relatively fragile type of meteorite that we think is actually very frequent in space but of which few pieces survive to reach the ground.

The carbonaceous chondrites are distinguished by containing a few percent of water and other volatile elements, indicating that the major masses of the asteroidal body from which they were broken were not greatly heated in their formation. Some people even suggest that they may be remnants of cometary nuclei that have lost their ices. This conjecture is highly controversial and most of us doubt its truth. R. N. Clayton, L. Grossman, and T. K. Mayeda showed that certain inclusions, or broken pieces, imbedded in the carbonaceous matrix of the Allende meteorite contain a dearth of the rare isotopes of oxygen with atomic weights of 17 and 18, as compared to the Earth, Moon, and most meteoritic samples. This component of nearly pure oxygen 16 is difficult to explain unless the oxygen 16 was made in a star or supernova and preserved in particles contributing part of the solar nebula. G. J. Wasserburg and his associates at California Institute of Technology's laboratory for studying lunar samples find evidence in some of these inclusions for an excess of magnesium 26, associated with a relatively high aluminum 26 concentration. The *chemical* reactions that occurred during formation of the meteorite would have permitted inclusion of aluminum but not much magnesium. Thus all the magnesium present now must have come from *nuclear* decay of aluminum 26 since the formation of the meteorite. As table 1 shows, aluminum 26, which decays to magnesium 26, has a half-life of only 740,000 years. The contribution of this short-lived isotope could have been made by a supernova that exploded not many million years before the beginning of the solar system. If the explosion occurred much earlier than this, all of the aluminum 26 would have decayed before formation of the meteorite. There is also evidence of special isotopes of xenon derived from extinct plutonium 244, which has a half-life of only 86 million years.

It now seems likely that the stellar incubator for the solar system was disturbed by a supernova, which exploded rather near in space and time to the fledgling solar system. The occurrence of supernovas in star-forming regions is quite frequent. The question remains as to whether the supernova may have helped precipitate the collapse of the interstellar cloud that made our system or even encouraged the Sun to accumulate our planetary system instead of a twin companion star.

FRED FRANKLIN

COMPUTER SIMULATIONS OF THE BEGINNING

The formation and evolution of the solar system certainly has gastronomic analogies, as Fred Whipple has shown. Indeed, I had thought of adapting my discussion to a Julia Child imitation—a sort of "Mastering the Art of Astronomical Cooking"—but I quickly decided that I could more easily discuss the subject in the terms of a very provincial and amateur galactic short-order cook featuring old, dusty, even radioactive freeze-dried food.

Fred Whipple's basic recipe also helps to limit the number of solar system models I might otherwise have needed to discuss. The reduced abundance of deuterium and lithium in the Sun relative to, say, the Earth means that we can relegate to lower probability models in which planetary material was somehow gathered from the Sun once it began to radiate by thermonuclear reactions. Instead we can concentrate on the model most astronomers find quite compelling: the formation of the Sun and planets from a collapsing and evolving cloud of what originally was interstellar dust and gas. Much of this topic has become more quantitative than could have been imagined not long ago. The application of physical principles and computer modeling replaced much speculation about solar system evolution.

We have seen stars—even the Sun—eject large amounts of material, and we have seen groups of stars, apparently formed as neighbors, separating from each other. What we have not seen with certainty, but whose existence we infer, are the processes of condensation and collapse of interstellar dust-gas clouds from a more nearly spherical distribution. One final disclaimer: we are attempting to unravel some 4.5 billion years of history from an assortment of scattered observations. From them, using physical principles, we can make predictions to be checked. However, what is taken to be proof and what is issued as a prediction can readily be inverted or even all scrambled together.

Let me start by calling attention to the presence of large, roughly spherical interstellar clouds. We can estimate their masses (approximately one hundred to one thousand times that of the Sun) and their densities (hundreds, or even thousands, of atoms and/or molecules per cubic centimeter). They have a gaseous component, which is actually most of the material and consists principally of hydrogen and helium, while the dust—some 1 percent of the total mass—consists of a wide assortment of submillimeter-sized particles, or grains, containing among other atoms those of carbon, iron, magnesium, silicon, calcium, sulfur, and oxygen, joined in a large array of molecules. These clouds are normally stable, in a pressure equilibrium with the more tenuous and hotter interstellar gas. Because all clouds are moving within the galaxy, they have the chance to undergo some change—possibly gravitational perturbations—that can upset the balance and initiate a slow collapse. The presence of a nearby supernova—the explosion of a rapidly evolving star much more massive than the Sun—would provide a more dramatic and effective mechanism to initiate collapse of an interstellar cloud. A supernova also produces grains as its expanding shell cools. Evidence that this may have occurred in or near our solar system is provided by the detection in meteorites of certain short-lived and now-extinct radioactivities produced in supernovas. (See discussion by Fred Whipple.) Once collapse starts, internal gravitation increasingly dominates any force that promotes expansion; the collapse can continue as long as energy is radiated into space. Because we know that large clouds contain density inhomogeneities, we can expect that, within a total cloud complex perhaps a thousand times more massive than the Sun, a number of individual condensations—fragments—will develop. Something very like this must have happened, because most stars, the Sun included, appear at least some time in their lives to have been members of clusters or associations. The fragmentation can continue until each individual entity develops, through gravitational attraction, a degree of central condensation that imposes some order on the surrounding dust and gas. What concerns us is the development of a fragment that has a few solar masses worth of material.

We can expect that most fragments will possess some degree of rotation. Without rotation, all material would collapse to a fragment's center to form a single star, and we are now certain that double and multiple star systems are much more common in space than single stars. As the collapse of a fragment continues, the amount of the rotation or spin increases and becomes better defined. Thus the system, at first roughly spherical, begins to flatten in a characteristic time of about 1 million years. This collapse and flattening, both predicted by computer modeling, represent the triumph of gravity over forces associated with pressure, random motions, and possible magnetic effects that try to maintain a distended cloud or solar nebula. There begins to emerge a giant pancake, extending many times farther than the present limits of the solar system (Pluto is forty times the Earth-Sun distance) with a large amorphous blueberry, the proto-Sun, that is beginning to be self-luminous at the center. One clear result is that the flattened, contracting disk rotates increasingly rapidly. This result has been known, at least in some approximation, for a long time. You know the effect illustrated in figure 1: as our friend with the dumbbells draws them closer to himself, he speeds up. The same principle, conservation of angular momentum, causes a large, slowly rotating

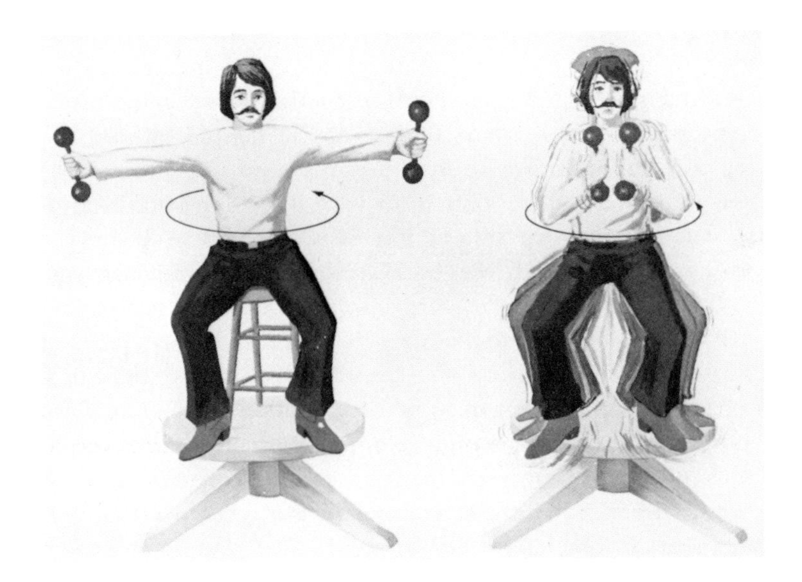

Figure 1
An example of change in spin rate with change in mass distribution. At any time, as long as our subject is unaffected by other forces, the product of the total mass × average distance from the spin axis × spin velocity will be a constant; that is, angular momentum is conserved. (From A. Baez, *The New College Physics,* W. H. Freeman, 1967.)

disk, or even a spherical cloud, when reduced in size by self-gravitation, to rotate more rapidly.

The fact that the Sun rotates so slowly—about once a month—means that most of the angular momentum in the solar system resides in planetary orbital motion. This fact had no easy explanation when these ideas were first introduced. And, for several generations, astronomers sought other explanations for the solar system's origin. We owe much of the recent revival of the nebula hypothesis in the last two decades to the power and scope of computer studies, for we can now follow in some detail the evolution of a flattened disk plus proto-Sun in which we can realistically include the effects of gas pressure, gravitational forces, and energy dissipation. We can also take account of the heating by the proto-Sun, whose incipient nuclear furnaces help to produce turbulent currents within the rotating disk around it. The results of several careful studies are in close agreement with each other and contain some surprises. One surprise is that a small fraction of the matter in the disk spirals outward from the center, while the bulk, some 90 percent, continues its inward progress. These studies show that some angular momentum is carried away by the outward moving stream—an outward-moving set of dumbbells to offset the inward-moving ones, thus conserving the angular momentum of the entire system and thereby permitting further collapse of the nebula without the inner portions speeding up to impossibly high velocities.

You may well object, This is all very well, but wouldn't the Sun, when the entire system has fully evolved, still show a shorter rotation period than once a month? The answer is yes, indeed; and observations of very young stars slightly more massive than the Sun indicate that they do spin rapidly, wth periods of approximately ten hours, and, further, that they are expelling a sizable fraction of their mass in a few million years. The mass shed by these stars carries angular momentum (a prime example of this phenomenon is the spectroscopically observed wind shown by the T-Tauri stars), causing such objects to slow down subsequent to formation. Therefore it is reasonable to expect similar behavior from the Sun. In fact, other arguments suggest that the Sun has lost a large fraction of its rotational angular momentum in the cosmologically short time of about a million years. A comforting recent observation is the discovery of a continuing solar wind, implying that the Sun's rotation is still decreasing, although now at a very slow rate.

Where does all this verbiage and computer-born insight leave us? Our contracting cloud now has a central and curious nucleus that because of infall of material begins to assume starlike properties. Its surface temperature is now a few thousand degrees, and so it begins to affect the disk in other than gravitational ways. The inner parts of the disk have temperatures slightly greater than a thousand degrees but because little solar radiation can penetrate radially into the fairly opaque disk, the temperature falls with increasing distance to the Sun, dropping to a few hundred degrees at a few hundred times the Sun's radius and then to only a few tens of degrees farther out. Within the disk-nebula the dust grains that were initially just pinpoints in size will grow to millimeter- or even centimeter-sized bodies as they fall to the plane of the disk, where they will also have a radial motion (figure 2). The gaseous component has flattened too, but because the molecules that constitute the gas are much less massive than the enlarged grains, the gas, even in the cooler regions, forms a thick envelope around the more flattened disk of dust.

At this point our solar system resembles a giant Saturn complete with ring, but this ring is not so flat and well ordered, and it is surrounded by more gas. Remember that the gas outweighs the dust by approximately one hundred to one. Suppose we continually added material to such a ring, much as would happen if material were continuing to condense onto it. There is a calculation we can do with a hand calculator and even picture the result. As mass is added, forces that try to produce clumpiness in the ring—self-gravitation between particles—begin to dominate the other effects, such as differential rotation, tidal

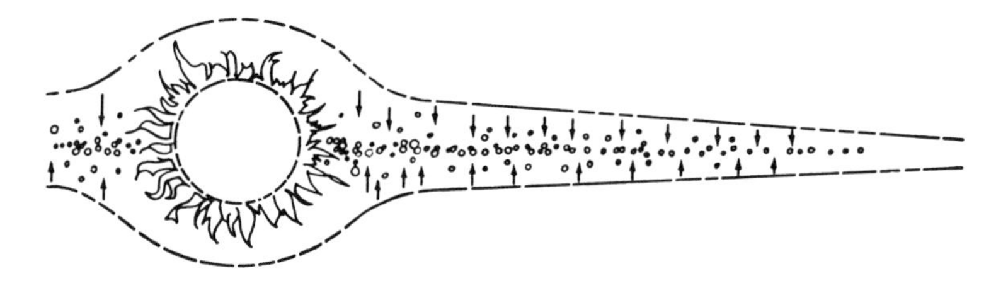

Figure 2
A schematic representation of dust grains falling under gravity to the mid-plane of the solar nebula. If sufficient mass is added, the thin sheet of dust that collects can become unstable and form roughly kilometer-sized bodies. (From J. A. Wood, *The Solar System,* Prentice-Hall, 1979.)

raising by the central mass, or random motions, that try to maintain the ring as a continuum. Once the space density of material exceeds a certain value, small gravitating condensations begin to appear. If you do not believe it, all you need do is double or maybe triple, the mass in the denser part of Saturn's ring and you will see some dramatic changes. Then suppose we grind up all the solid material in the inner four planets, Mercury to Mars, and spread it out as a thin disk. We would find that this disk had a surface density of approximately 10 grams per square centimeter; or if its thickness was about 1 kilometer, then its space density would be one ten-thousandth gram per cubic centimeter. That is, not very much, but remember we began with a cloud of only a few thousand molecules per cubic centimeter (of which the dust was only a small part) or more than a million billion times smaller. The question is, Would this primordial dust layer, composed of grains of roughly millimeter size, be stable, or would some clumpiness result? The answer is that such a smooth, continuous sheet is quite certainly unstable and would develop in the course of only a few hundred years into an aggregate to kilometer-sized, self-gravitating bodies—not planets, admittedly, but what we might call planetesimals. We have known about this result only for some eight years. Had the thin dust layer possessed a density some one hundred times greater, the condensations might have been close to planetary size and we could stop here. It seems quite possible that the density was larger, but not one hundred times larger, in the vicinity of the major planets (Jupiter and Saturn)—in part because the lower temperature in that region would permit the condensation of some icy material in addition to the minerals spawned in the dust.

Another basic fact favors the growth of large objects in regions not as near to the Sun as is the Earth. Consider the hypothetical case of a large, loose clump trying to organize itself. Its self-gravitation must dominate the tidal forces of the Sun that are trying to shear the condensation apart. Because the force associated with tidal shearing decreases in proportion to the cube of the distance of the Sun, successful condensation is more likely to occur at larger distances. Thus at five times the Earth's distance (Jupiter) or ten times (Saturn) we expect to find larger planetesimals. We also suspect that they formed sooner.

To summarize, we now have a developing Sun surrounded

by literally billions of kilometer-sized bodies. In actuality these bodies vary widely in size and composition, as I have implied. In the course of time, there are bound to be encounters and collisions between the members of so abundant a supply of bodies. Here is a good problem for a modern computer. We introduce a dominant central body, the Sun, and an array of kilometer-sized bodies surrounded by an envelope of hydrogen and helium and follow the subsequent motions and possible growth.

Figures 3 and 4 show the results from a first attempt at the problem, made some ten years ago. In this particular case, small, already formed planetesimals were injected randomly and sequentially into the condensing disk of dust and gas, and the computer was asked to run until all the remaining dust and gas was exhausted. What we see is the result of a grand sweeping (*accretion* is the usual word) of small bodies to form larger ones. A growing planetesimal gradually accretes a wide swath of neighboring material. And if it grows to a mass a few times that of the Earth, it can also retain gravitationally the light gases, hydrogen, and helium. Naturally capture and retention of gas are easier far from the Sun, where the temperature is lower.

Now let us consider a new generation of more exact computer models. From ballistic experiments, we have an idea of how to model collisions between objects of different compositions and crushing strengths. If two bodies collide with high velocity, they are likely to fragment but if their relative velocity is low, the two will stick and become one, although some collisional debris will be produced. The computer can be asked to model and keep track of all bodies. Two objects that almost collide will have their eccentricities increased, while a small planetesimal already in an eccentric orbit will experience some drag from the uncondensed dust and gas, a process that tends to circularize and reduce its orbit. (The general shape of all the orbits is elliptical. The semimajor axis of the orbit refers to half the length of the longest axis of the ellipse; the eccentricity of the orbit refers to the degree of circularity of the ellipse, ranging from 0 for a perfect circle to 1 for an ellipse flattened to a line.) These second-generation results show that from a wide variety of initial conditions and assumptions, near lunar-sized (approximately 1000 kilometer) bodies can accrete in a time very much shorter than the solar system age of 4.5 billion years. Specifically, from a million million bodies, each approximately kilometer-sized

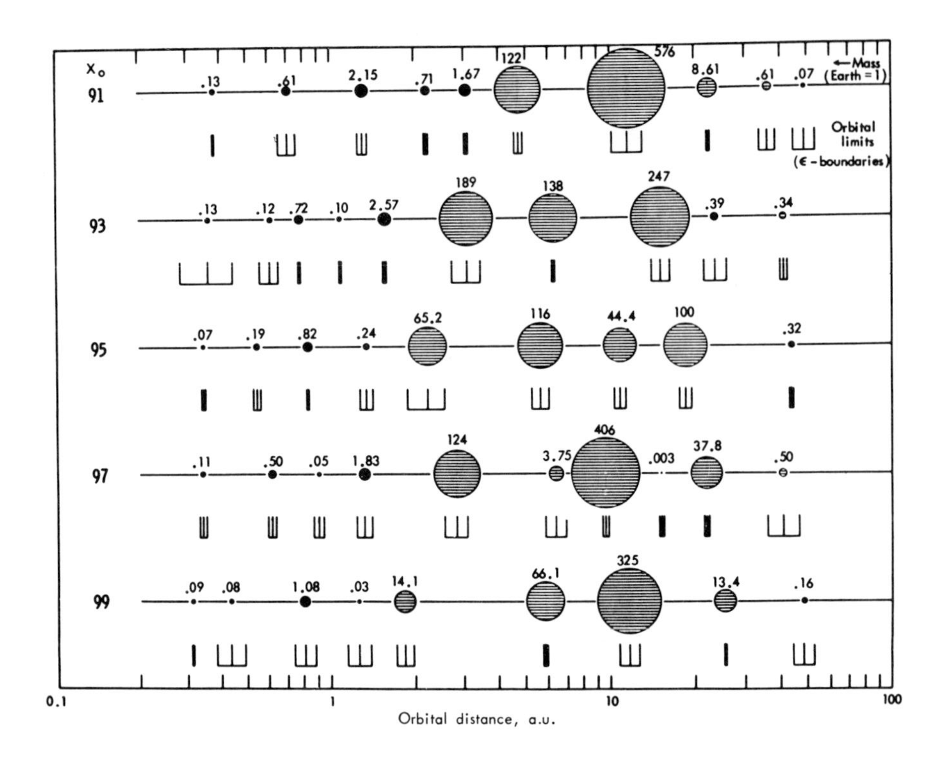

X_o
91
93
95
97
99
←Mass (Earth = 1)
Orbital limits
(ε - boundaries)
.13 .61 2.15 .71 1.67 122 576 8.61 .61 .07
.13 .12 .72 .10 2.57 189 138 247 .39 .34
.07 .19 .82 .24 65.2 116 44.4 100 .32
.11 .50 .05 1.83 124 3.75 406 .003 37.8 .50
.09 .08 1.08 .03 14.1 66.1 325 13.4 .16
0.1
1
10
100
Orbital distance, a.u.

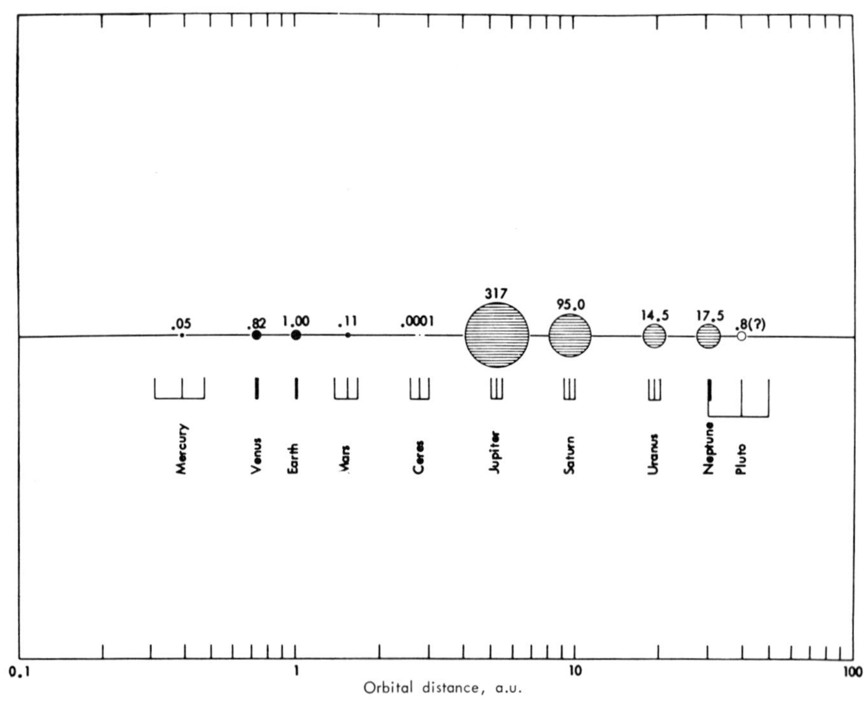

.05 .82 1.00 .11 .0001 317 95.0 14.5 17.5 .8(?)
Mercury
Venus
Earth
Mars
Ceres
Jupiter
Saturn
Uranus
Neptune
Pluto
0.1
1
10
100
Orbital distance, a.u.

initially, lunar-sized objects appear in about ten thousand years. Multiply the initial number by one hundred and they appear in only about one hundred years.

This exciting and physically realistic result prompted an even more exact calculation, just finished, that followed in detail the evolution of one hundred Moon-like bodies in an effort to produce the terrestrial planets. This third-generation calculation produces what is almost safe to call the inner solar system if one important prescription is observed; the initial bodies must move, and continue to move, in somewhat eccentric orbits. Too low an initial eccentricity or too circular a set of orbits produces too many planets that are themselves too small. It seems unlikely that longer computer runs would alter this conclusion; these bodies do not experience sufficient gravitational or collisional interaction to continue growth. If the eccentricities are too large, then too few planets result. A value of eccentricity about 0.10 seems about right.

I would like to provide a clear justification of exactly this value, but at present that is difficult to do; however, the following comments may be helpful. If Jupiter did form earlier and acquired its mass of some three hundred times the Earth's value, it could, in addition to accreting bodies for its own growth, efficiently spray other objects all over the solar system. In fact, the intense cratering seen on almost all planetary and satellite surfaces requires the energetic impacts that such gravitational scattering would produce. Thus Jovian perturbations, acting directly or by producing this shower of scattered planetesimals, might keep the accreting bodies in the inner solar system from nestling comfortably into circular orbits. In some sense it is heartening to know that the asteroids—the small (about 100-

Figures 3 and 4
A first-generation calculation (after S. H. Dole) of the growth of planetary models by accretion in a gas and dust medium. Numbers above each model planet indicate mass (Earth = 1) after growth ceases, and trident marks measure final orbital eccentricities. Orbital distances from the Sun are shown on the bottom axis and given in AU (astronomical units), where 1 AU is the distance from the Sun to Earth. Hatching indicates that a body has acquired sufficient mass to retain gravitationally the light gases, hydrogen and helium, at the local temperature. The actual solar system is included as figure 4. Although these calculations have some relevance for the formation of the terrestrial planets, recent work argues that the major planets are more likely to have originated from density inhomogeneities in the solar nebula.

kilometer) bodies lying near Jupiter—have average eccentricities close to 0.15, for they also would have been affected by assorted perturbations. Our present thoughts bend in the direction of using Jupiter to trigger the formation of the inner solar system, much as the supernova was called upon for the collapse of interstellar clouds.

In the case of the minor planets in the asteroidal belt, the trigger seems to have been sufficiently violent to influence their evolution. The quoted mean eccentricity of these bodies is a factor of about two greater than the value $e_{max} = 0.15$ (figure 5) that was most appropriate for forming terrestrial planets. It is also clear that there is only a small amount of material now

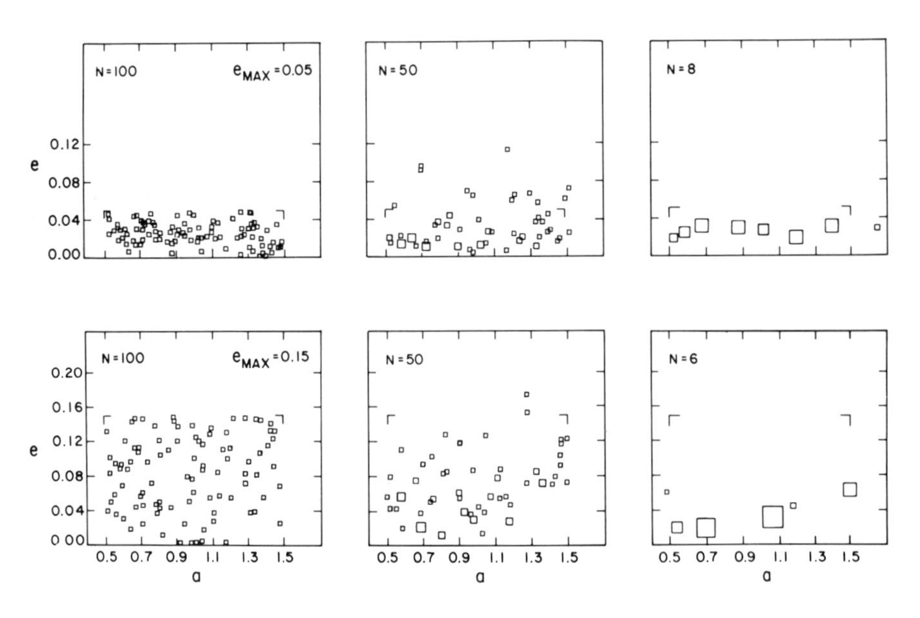

Figure 5
Two examples of the accretion of lunar-sized planetesimals in the attempt to simulate the terrestrial planets. Initially N = 100 bodies are placed, with random semimajor axes (indicated by *a* on the horizontal axis) and eccentricities (indicated by *e* on the vertical axis), chosen within the indicated limits. The computer follows the details of their subsequent gravitational and collisional evolution, ending the calculation when mutual interactions fall below a specified level. Longer runs probably would have reduced slightly the final number of bodies in the $e_{max} = 0.15$ example, making it resemble still more closely the actual case. N's refer to the number of remaining objects and the area of the squares measures their masses. (After L. P. Cox, Ph.D. dissertation, Massachusetts Institute of Technology, 1978)

remaining in the asteroidal belt—less than one-twentieth of a lunar mass. This translates into a very low collision frequency among the present-day population; each asteroid probably collides once every billion years. Thus the present (and past) accretion rate of asteroids is very slow, both as a consequence of the infrequency of collisions and of their high relative velocities. Some accretion must have taken place in the past, however, because some 75 percent of the total mass of all minor planets resides in the four largest. In summary, we now are quite certain that Jovian perturbations are ultimately responsible for the large asteroidal mean eccentricities and also for the reduced number density of the population, probably converting one-time asteroids into impact craters on the larger members of the solar system.

This picture can now be used to make a few predictions. The first is a sort of anticlimatic prediction. A long-established relation is Bode's law, which quantitatively shows a regularity in the spacing of planetary orbits. It is not an overstatement to say that all computer-simulated models also exhibit a Bode's Law (figure 6). It seems as though this relationship is a characteristic of the accumulation of bodies in a gravitational field, perhaps reflecting the simple fact that the successive planets evolve (by devouring their neighbors) to such a separation that they no longer greatly perturb each other. We suspect, therefore, that the presence of a Bode relation proclaims that a system has essentially completed its collisional and gravitational development. Its very universality means that it provides no special information concerning initial conditions or evolutionary details.

The second prediction concerns the amount and direction of the spin of planetesimals and planets. If an accreting body encounters material moving in strictly circular orbits, it would gradually acquire a spin in the direction opposite to its rotation about the Sun. But if the body meets material in slightly eccentric orbits, the arrival pattern is very different. Although I cannot provide a simple and direct explanation, computer studies lead to the clear result that spin in the same direction as revolution about the Sun must develop. Again we find the need for noncircular orbits. There is one rather exciting prediction of this modeling: the characteristic rotation period of an accreting body should be in the neighborhood of thirty hours and should be independent of the body's mass. That is reasonably heart-

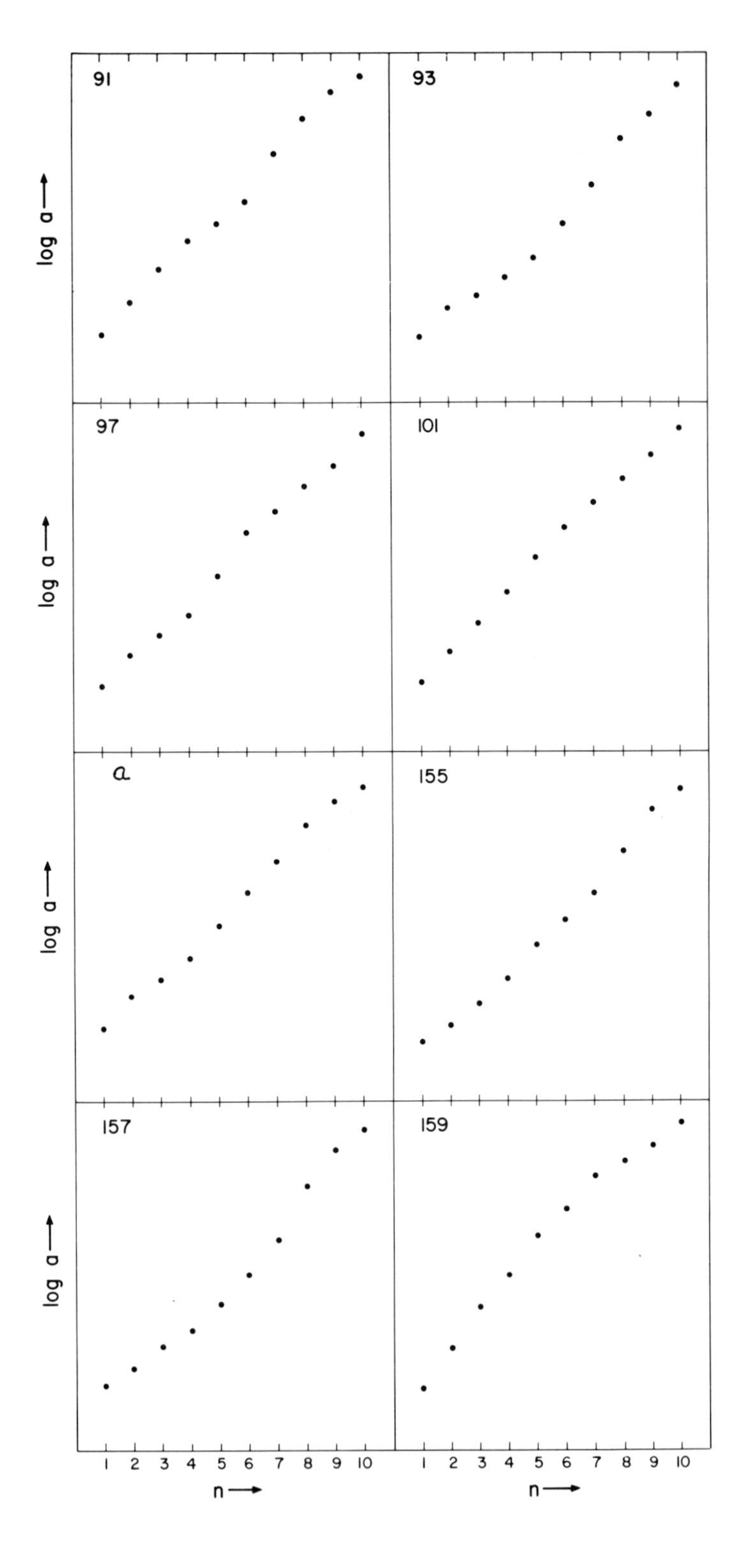
91
93
97
101
a
155
157
159
log a
n
1 2 3 4 5 6 7 8 9 10

Table 1
Period and obliquity of planets

Planet	Rotation Period(Hours)	Obliquity (Degrees)	m_f/m_p
Earth	24.0	23.5	0.001
Mars	24.6	25.2	0.002
Jupiter	9.9	3.1	0.0003
Saturn	10.7	26.7	0.04
Uranus	16.2	98.0	0.07
Neptune	11	29	0.007

Note: Columns 2 and 3 give the rotation period and obliquity, the angle between a planet's equatorial and orbital planes. (Planetary perturbations cause values of the latter to change, especially in the case of Mars, whose obliquity varies by $\sim\pm 10°$ from the average value.) The final column gives, in terms of the planet's mass, m_p, a characteristic value, m_f, for the mass of impacting bodies that are required to produce the observed obliquity from an assumed initial value of 0°.

ening even though Jupiter's period is just under ten hours and the tiny asteroids, some tens of kilometers in diameter, show periods from five hours to several days.

The third prediction is that the spin axis of a planet should be more or less perpendicular to its orbital plane because most of the material to be accreted would lie, on average, in the planet's orbit. For every planet, the final value of this inclination angle will be determined by the collisions of the last few large fragments that went to make up the planet. If the last large planetesimal happened to strike near a pole, then the axis would be turned over somewhat. More than somewhat is exactly what we suspect happened in the case of Uranus, but notice that the mass of the last colliding body is a reasonable value—still only a few percent of the planet's total mass. (See table 1.)

The fourth prediction is one Fred Whipple discusses: the

Figure 6
Bode's Law in the actual ("a") solar system and in several of Dole's computer-generated models. The logarithm of a planet's semimajor axis, a, is plotted vertically against its number in order of increasing solar distance. (Other, equivalent representations of the law are possible.) Mean distance of the asteroidal belt (2.8 AU) is included as a fifth planet, although it is very unlikely all the asteroids were once collected together as a single object.

mean densities of planets should decrease with solar distance. Broadly speaking, the terrestrial planets have high densities; the outer planets, even when allowance is made for their thick atmospheres, show lower ones. Near Mercury's orbit, we have argued that the temperature was sufficiently high (greater than 1000 K) so that only iron and certain other chiefly iron-bearing minerals could exist as grains rather than as vapors. For the outer and cooler planets, iron minerals are present, but so are the less dense icy volatiles in great abundance. This pattern is repeated in the Jovian satellite system, where the innermost large (Galilean) satellite, Io, has a density about twice that of the outermost large satellite, Callisto. Thus we are tempted to speculate that our outline for a possible history of the solar system was repeated on a reduced scale by Jupiter. This viewpoint is reinforced by the recent discovery that both Jupiter and Saturn radiate about twice the energy each receives from the Sun. Jupiter was in fact almost a star; had it gathered maybe some twenty times its present mass, it would have been one. The presence of a residual Jovian ring is also suggestive of a general pattern.

Before I make a leap of faith and issue a fifth prediction relating to satellite formation, I must point out that Saturn, also with a ring but over three times less massive than Jupiter, follows a different behavior. Its satellites show very little variation in density with increasing distance from the planet. The presence in the Jovian, Saturnian, and even Uranian systems of regular satellites (bodies whose orbits have small inclinations and eccentricities) and rings is not in easy accord with the view that these planets grew only by accreting planetesimals—the mechanism that works well to produce the inner planets. When it comes to predicting the presence or absence of a ring and large numbers of satellites, growth by accretion clearly forecasts absence. However, if we are willing to make the reasonable assumption that the solar nebula could develop and sustain a few density fluctuations (in addition to its central condensation), then the major planets would grow directly from the surrounding disk material. Instead of developing a core massive enough to collect and retain gaseous material, this view proposes that the core condensed out of a large region of greater than average density. The density of material required to form self-gravitating condensations is still low; to form Jupiter, the necessary value lies in the range 10^{-8} to 10^{-11} grams per cubic centimeter, depending on the details of the Sun's influence.

Computer modeling of this process shows that proto-Jupiter and Saturn were extensive flattened disks of material, many hundred times their present radii. Contraction to form a central core again occurs—over a time of about 1 million years (Jupiter) and 10 million years (Saturn)—with a residual equatorial disk that transmits angular momentum outward. (Because contraction time varies with the square of the protoplanetary mass, major planets with masses much smaller than those of Uranus and Neptune could not appear in the allotted time of a few billion years.) In the contracting stage, enough gravitational energy is liberated as heat to prevent the deposition of icy material in the disk. Once pressure in the planetary core greatly slowed further contraction, the rate of energy production fell by a factor of about one hundred (to its present small value), allowing ices to condense sequentially on small bodies within the disk to form the satellites we see and continue to discover. In this picture, the relatively high mean density of the two inner Galilean satellites, Io and Europa, is in part the consequence of tidal interaction with the primary. This process also leads to energy dissipation and heating, which in turn evaporates the low-density volatiles that might otherwise have been deposited. We suspect, too, that Jupiter, with its relatively high mass, radiated enough heat during its contraction to reduce the chance for ice to condense on nearby satellites.

The formation of the major planets by means other than direct accretion of planetesimals alone has the appealing feature that it could predate the process that produced the terrestrial planets. Thus we have a mechanism that, in addition to allowing extensive satellite formation, provides at the right time the necessary mass to generate eccentric orbits for the planetesimals in the inner solar system.

A last prediction: we do not exist!—well, almost did not exist. Another surprise from the computer simulations is that, in addition to a flattened disk, most collapsing fragments produce not a simple central condensation but a thick bar of material. This bar is not stable and appears likely to separate into (usually) two roughly spherical bodies that orbit one another—a double star. Except for the fact that Earth happens to orbit a single star, this is a nice result: most stars are double. Under barely

Figure 7
The character of the protostar-nebula system formed by contraction of a cloud fragment would depend upon the angular momentum of the latter. A fragment with little angular momentum would produce a large protostar and minor nebula (A). More angular momentum would produce a more spun-out system (B), in which the nebula has increased prominence. A very rapidly rotating cloud fragment would produce a binary or multiple protostar system upon contraction (C). (From J. A. Wood, *The Solar System,* Prentice-Hall, 1979)

possible conditions the two protostars might coalesce to form one; more often they would not. In any event, a double star is a more likely result than a solar system (figure 7). For example, if Jupiter had a mass some twenty times its present value, it would have shown its gratitude not merely by setting off nuclear reactions but also by ejecting most of the other planets from the solar system. More to the point, Jupiter probably would have prevented any additional planetary formation in the first place. Only if the initial fragment had a low angular momentum, within fairly small limits, is our present configuration possible. Thus the number of life-bearing planets throughout the galaxies may be many fewer than past speculations have proposed.

I have reviewed a number of separate and simplified parts of a puzzle. Now it is time to look forward to a more rigorous calculation that incorporates all the effects I have mentioned, such as collapse, angular momentum transport, growth of density fluctuations possibly associated with the barlike instability, accretion, and the solar wind in one grand synthesis.

Further Reading

Dole, S. H. *Habitable Planets for Man.* New York: American Elsevier, 1970.

Lewis, J. S. "The Chemistry of the Solar System." *Scientific American* (March 1974).

The Solar System. San Francisco: W. H. Freeman, 1975.

Ward, William R. "The Formation of the Solar System." In *Frontiers of Astrophysics,* edited by E. H. Avrett. Cambridge: Harvard University Press, 1976.

Wetherill, G. W. "The Formation of the Earth from Planetesimals." *Scientific American* (June 1981).

Whipple, Fred. *Orbiting the Sun.* Cambridge: Harvard University Press, 1981.

Wood, J. A. *The Solar System.* Englewood Cliffs: Prentice-Hall, 1979.

CHAPTER 4

THE PUZZLE OF THE SUN'S HOT CORONA

ROBERT W. NOYES

A MATTER OF DEGREES

The chapters by Kenneth Brecher and Owen Gingerich trace the history of the development of the theory of gravity through several attempts to predict and later to understand planetary motions in the solar system, culminating in Einstein's General Theory of Relativity. Einstein's theory, based on only a few fundamental postulates, was a grand unification, bringing our ideas of space, time, gravitation, and cosmology into a single framework. So compelling was that framework that upon its conception Einstein knew it had to be correct. (Readers will recall Einstein's lack of expressed delight at the news of the successful measurement of the bending of starlight at the solar eclipse of 1919. If the results had not verified the theory, he said, "Then I would have been sorry for the dear Lord; the theory *is* correct.") In such grand theoretical schemes, observation is almost secondary; it serves simply to verify the theory's validity.

Not all theoretical astrophysics deals with such grand, simplifying descriptions of reality. Some of the most fascinating phenomena in the universe involve extremely complex interactions of many individually simple entities. For example, although the problem of gravitational interaction between two bodies is solved, if a large number of gravitating bodies, such as the stars in our galaxy, interact, their detailed motions cannot be predicted by even the most complicated theoretical formula. Rather such motions must be painstakingly calculated, step by step, numerically summing all the pairs of gravitational interactions at each instant of time. A second example is embodied in the puzzle of the Sun's hot corona. No grand unifying scheme will tell us a priori that stars like the Sun must be surrounded by extended atmospheric shells, called coronas, or that these must be heated to a particular temperature. The rich brew of coronal physics has been cooked from a very complex recipe. The basic ingredients (gravitation, magnetic fields, radiation, nuclear processes) are each understood in terms of fundamental theory; however, the flavor of the final concoction can be discerned only by elaborate tasting—in other words, observation. Here the roles of observation and theory are closely intertwined. Observational sampling of the result gives suggestions for theoretical descriptions of the recipe, and various theoretical recipes are subject to observational taste tests. The goal is to derive a final recipe that produces the observed brew. However, the recipe is so complex that it could not have been developed in the absence of observational narrowing of possibilities.

Because of the inseparable interplay of observation and theory in problems such as the heating of the solar corona, Robert Rosner and I have elected not to attempt a separate description of observational approaches, followed by a similar description of theoretical approaches. Rather we shall take a more historical approach, in which I discuss the development and interplay of observational and theoretical studies of the corona up to the recent past, and Robert Rosner discusses the present theoretical understanding of the corona puzzle and how this current recipe suggests that observation and theory may progress in the future.

Most of the time the Sun appears to us as a perfectly round disk in the sky. When we look at it through smoked glass or the dimming sky of sunset, we see that this disk has a very sharp edge; in fact, even with the largest telescopes, the edge of the Sun seems perfectly sharp, with only empty space above it. Occasionally, however, during a solar eclipse, the Moon completely covers the bright disk, and on such occasions a faint pearly-white halo is seen surrounding the Sun; this halo is known as the solar corona (figure 1). During these eclipses, we can see that the coronal halo is highly structured, consisting of bright and dark regions, with even brighter areas, called streamers, appearing to shoot out radially from the Sun for great distances. In some eclipses streamers are seen to extend outward as far as 20 solar radii, some 9 million miles, or 10 percent of the distance from the Sun to the Earth. Some eclipses show a corona concentrated mainly toward the equator of the Sun, with almost no coronal emission from above the poles (figure 1); others show emission all around the Sun (figure 2). Thus the corona not only has structue, but also the structure changes with time.

Just what is the solar corona, and why is it there? Although the corona has been seen for untold centuries, such questions could not be seriously asked before astronomers had developed observational and theoretical tools to study it more carefully.

Figure 1
Solar eclipse showing minimal coronal structure and activity concentrated at the solar equator. (Photograph from Smithsonian Astrophysical Observatory)

With the development of the spectrograph, which measures the intensity of light of different colors, it was found that the coronal spectrum resembles very closely the spectrum of the Sun's disk itself, suggesting that the light we see is simply normal sunlight, scattered off tiny particles that surround the Sun. (The light of the corona was also found to be polarized, as would be expected theoretically from the scattering of Sun's light from small particles swarming around it.) The corona seemed to be an extended region of gaseous matter or, in other words, an atmosphere, surrounding the Sun.

On the surface, this sounds very reasonable. The Earth has an atmosphere; why shouldn't the Sun? A major problem with this analogy appears to be the vertical extent of the Sun's atmosphere. The Earth's atmosphere becomes rapidly rarefied

Figure 2
Solar eclipse with extended corona around the Sun in all directions. (Photograph from Harvard College Observatory)

with increasing altitude, so that at an altitude of 5 miles above sea level the air pressure is only one-half as great as at sea level; at twice that altitude it is one-fourth as great, and so on. Yet the atmospheric pressure in the Sun's corona drops off much more slowly with height, so that one would have to climb some 50,000 miles before the pressure dropped to half of its value just above the surface. This is true even though the force of solar gravity, which should bind the Sun's atmosphere tightly to its surface, is some thirty times as great as the Earth's gravity. So if the solar corona is an atmosphere around the Sun, why is it so extended in altitude? What holds it up?

One way to hold an atmosphere up would be to heat it to a high temperature. Then the pressure of the atmospheric gases would be greater and they would expand upward. It can be shown theoretically that the extent in height of any atmosphere, whether the Earth's or the Sun's, is directly proportional to its temperature. But the Sun's corona, or atmosphere, could be held up this way only if it were very hot—at a temperature over 1 million °C.

This is actually the current explanation for the extended solar corona, but in the early twentieth century when astronomers were first grappling with the problem of the solar corona, they refused to accept this explanation. The problem was that although the core of the Sun was known to have temperatures as high as and in fact much higher than 1 million °C, the surface was relatively cold—only about 6000 °C. It was believed that energy always flowed from hot to cold regions, in agreement with both the Second Law of Thermodynamics and common experience. Thus the steady decrease of temperature from core to surface was in keeping with the flow of energy from the core, where it was created by nuclear reactions, to the surface, where it then was radiated into space and to the Earth. But what could cause the temperature to rise again to more than 1 million °C in the corona? If the corona is so hot, something must continually be heating it, for otherwise it would gradually cool off, transfering its heat back to the cooler solar surface or outward into cold outer space. It was difficult for theorists to see how a continuous flow of energy could come from the Sun itself, for the energy would have to flow uphill from cold to hot, in violation of the Second Law of Thermodynamics.

This was, and still is, the puzzle of the hot corona. In the 1930s when the puzzle was first posed more or less in this form, attempts were made to claim that in spite of its great extent in height, the corona was, in fact, not hot, so there was no puzzle. Thus E. A. Milne developed a theory that the corona was held up by radiation pressure—the weak force exerted by sunlight on matter in the Sun's corona when its particles absorb or scatter the sunlight. A clever idea, but radiation pressure failed to be a strong enough force when put to a quantitative test. In spite of that failure, there seemed no direct proof that the corona was hot, and no one wanted to be guilty of transgressing the Second Law of Thermodynamics.

Evidence that the corona was hot, however, kept piling up. In the early 1930s, a strong piece of evidence turned up in data that had actually been obtained over fifty years earlier, in the first eclipse expedition to obtain a *spectrum* of the solar corona. When a glass prism was inserted into the light path of an eclipse telescope, images were recorded showing the coronal light bro-

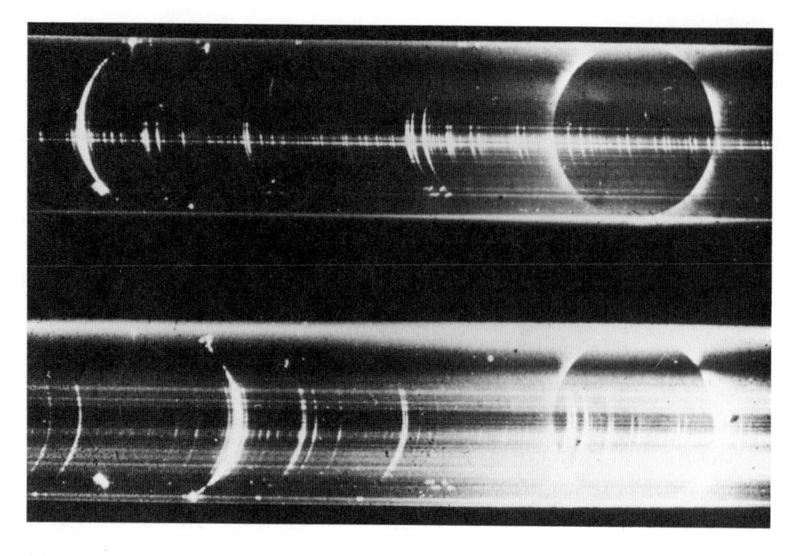

Figure 3
Flash spectrum of the Sun during solar eclipse. (Photograph from Harvard College Observatory)

ken up into its component colors. Figure 3 depicts such a spectrum, showing arcs of the bright low corona above the Moon's edge just after totality began. Each arc corresponds to light emitted by a different element. The bright arc on the left is due to hydrogen, the most abundant element in the Sun, and most other emission arcs may be identified with spectra of common elements observable in the terrestrial laboratory. The complete ring of emission near the right of figure 3, however, eluded identification. It was clearly the radiation of some element emitted so high in the corona that its emission formed a complete ring in the spectrum. However, its emission could be produced by no earthly chemical, so in despair it was named coronium and thought to be a rare new chemical element found only in the corona.

Finally, in the early 1930s it was noticed that the wavelength of the coronium line was just what one would expect from the very common element iron if thirteen of its electrons were knocked off. To knock off so many electrons, however, requires very high energy collisions, which could occur in the corona only if it were exceedingly hot. Many scientists were skeptical. One famous astrophysicist stated that it must just be a coincidence because otherwise the solar corona would have to have the impossible temperature of more than 1 million °C.

But, soon it was discovered that a number of other previously unidentified coronal emissions occurred at wavelengths corresponding to those predicted for common elements like iron or calcium heated to temperatures of 1 million °C or more, and it rapidly became clear that there was no explanation for the totality of these emissions other than that the corona really was extremely hot. Other types of evidence eventually became irrefutable, so the focus switched to asking not whether but why the corona is so hot. As we leave the first question behind, however, let us note that it represented just one of the multitude of blind alleys that astronomers traverse in their search for explanations and that, characteristically, it was the weight of observations, rather than theory, that finally forced the abandonment of speculation about a cold corona.

Although recognizing that the corona was hot, many astronomers were reluctant to accept the notion that the energy to maintain its high temperature flowed from the much cooler surface into the hot corona. Thus the eminent English astronomer Fred Hoyle advanced the hypothesis that the corona was heated by meteoritic infall—dust grains falling into the Sun, each one of which converts its kinetic energy of infall into a tiny bit of heat. Again, when put to a quantitative test, the theory failed, so astronomers were finally forced to confront the unpleasant task of getting energy from the much cooler solar surface into the hot corona.

As has frequently happened in astronomy, once theory was driven by observations into what was previously considered an untenable position (in this case an apparent violation of the Second Law of Thermodynamics), a way out was found. Although it is true that heat energy flows only from hot to cold (by thermal conduction or radiation), other forms of energy appear in the solar surface that do not have such a restriction.

Two such forms of energy underlie the two most obvious surface features on the Sun. One of these features is sunspots, dark vortices where strong magnetic fields stop the outward flow of heat energy and thus cause the surface to cool and become dark (figure 4). The other feature is granulation (figure 5), the seething convective eddies of rising and descending turbulent motions in the outer layers of the Sun. Each feature is

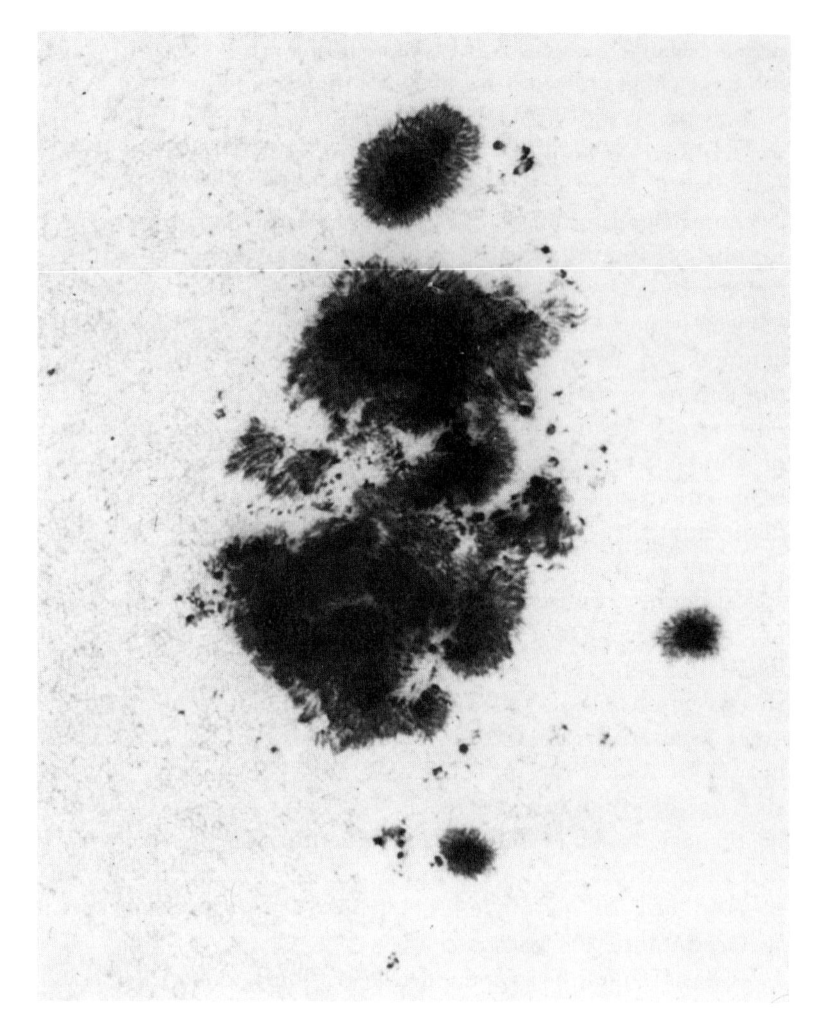

Figure 4
A large sunspot group photographed on May 17, 1951. (Photograph from Mount Wilson and Las Campanas Observatories, Carnegie Institution of Washington)

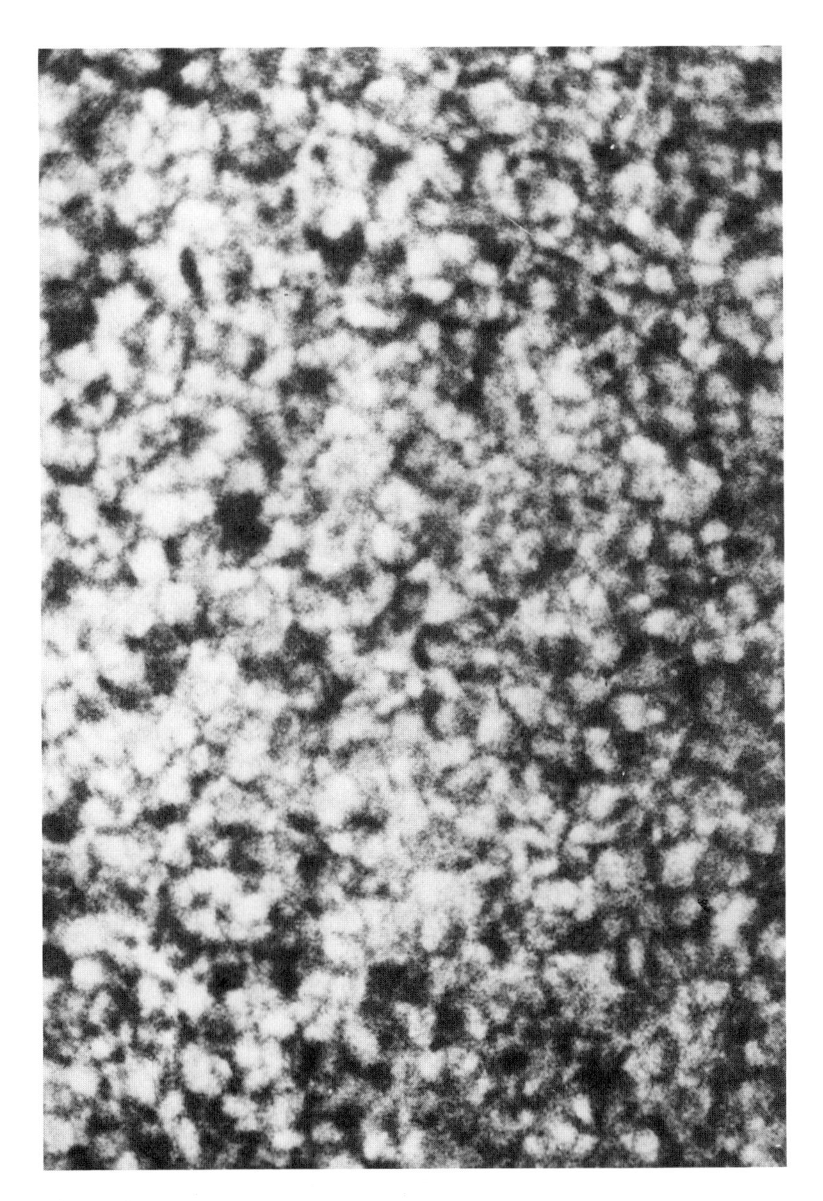

Figure 5
Direct photograph of photospheric granulation taken from the 12-inch balloon-borne telescope of Project Stratoscope. (Photograph from Project Stratoscope of Princeton University, supported by NASA, NSF, and ONR)

proof of the importance of a different type of energy: sunspots reveal magnetic energy strong enough to halt convective energy flow; the granulation reveals the kinetic energy of its turbulent motions. Either magnetic or kinetic energy can propagate from one place to another with little regard to which is hotter, so in principle either could provide a solution to the puzzle of the hot corona.

In 1946 the astronomers Martin Schwarzschild and Ludwig Biermann independently pointed out that the enormous kinetic energy in the granulation must generate violent sound waves. Could we but hear the Sun, we would be deafened by the noise of these turbulent motions, just as we are by the turbulent exhaust from a jet engine. Furthermore, it was reasoned, these sound waves would propagate up into the corona, in whose rarefied atmosphere they would strengthen and turn into shock waves—colossal sonic booms. Surely these sonic booms, generated by the granulation and amplified and dissipated in the corona, must be the source of the corona's high temperature.

This suggestion of shock wave, or acoustic, heating appears to have led astronomers down a second blind alley, this one so long and twisting that only recently have they emerged. (I say "appears to have led" because not all astronomers agree today that it is a blind alley, although almost all agree that this explanation cannot be the whole story.) The length of time—nearly thirty years—spent exploring this possibility testifies to the fact that once a simple and right-sounding theoretical explanation is suggested, the search for alternate suggestions becomes much less intense, and instead attention turns to working out details of the proposed explanation. Astronomers happily explored this blind alley until dragged out as if by the scruffs of their necks; and once again the agent that dragged them was new observations.

In fairness to the two astronomers who originally suggested shock wave heating, they probably had no thought that their suggestion would be so readily and thoroughly accepted; however, events conspired to make the suggestion particularly tantalizing. One event was the development of theoretical understanding of how turbulence generates shock waves. Analysis of data on jet engines showed that the shock generation rate increased very rapidly as the speed of the turbulent motions increased. When the theory was applied to the Sun, it gave very plausible predictions of heating rates. (M. J. Lighthill, the originator of the theory of production of shock waves by turbulent motions, later expressed considerable surprise that astronomers had picked up this theory for jet engines and applied it almost without alteration to the surface of a star.)

A second event conspiring to make the acoustic heating theory accepted almost without question was the observational discovery in 1960 that the surface motions of the Sun have a strong regular oscillatory component, with a period of five minutes. If we could hear the sound waves produced by the granulation, they would include a nearly pure tone, as well as the noise of the more random fluctuations. This pure tone in fact is a pitch far below what the ear can hear (thirteen octaves below the lowest note on the piano keyboard), but nevertheless it was an additional candidate for heating the corona by sound waves. Throughout the 1960s, numerous theoretical calculations were performed in an attempt to match the propagation of such waves with the temperature structure (by then measured in some detail) of the corona.

In retrospect it all seems classically naive: two phenomena were observed more or less at the same time—the turbulent motions of the granulation and the high temperature of the corona— and it was deduced that one caused the other. It would scarcely be different if a visitor from outer space were to observe two true facts—that much of the surface of the Earth (the oceans) experiences turbulent motions and waves and that the outer atmosphere of the Earth (the exosphere) is heated to the surprisingly hot temperature of more than 1000 K—and draw the totally erroneous conclusion that the second was caused by the first.

Until the 1960s virtually all observations of the corona were made from the ground, in visible light. Unfortunately, in visible light the faint coronal radiation is outshone a million-fold by the tremendous brightness of the solar surface and is observable only when the Moon (or an artificial occulting disk in a coronagraph) blots the brilliant light of the Sun's surface. However, there are much more favorable emissions in which to view the corona. Because the corona is so hot, most of its radiation is produced in the form of high-energy radiation, such as x-rays and extreme ultraviolet radiation. These do not have to compete with radiation from the surface of the Sun, which is relatively much cooler and emits almost no high-energy radiation. Also, the short-wavelength x-ray and far-ultraviolet emissions convey unique information about the coronal conditions that is not even contained in the feeble visible radiation.

Because the Earth's atmosphere is opaque to extreme ultraviolet and x-ray radiation (fortunately for our health), observations of the Sun's corona at these wavelengths had to await the Space Age. But now a long series of space missions, culminating in the extended series of solar observations aboard Skylab in 1973–1974, have given us a new view of the solar corona.

This view is illustrated in figure 6, one of many Skylab images of the Sun in x-rays. The brightest areas in the figure are regions that emit strongly in x-rays; this requires that they have a temperature of at least several million degrees centigrade and that they be dense enough to emit efficiently. Conversely dark areas are either too cold, too rarefied, or both, to emit x-rays efficiently. In photographs of this sort scientists saw for the first time how the solar corona appears when one looks straight

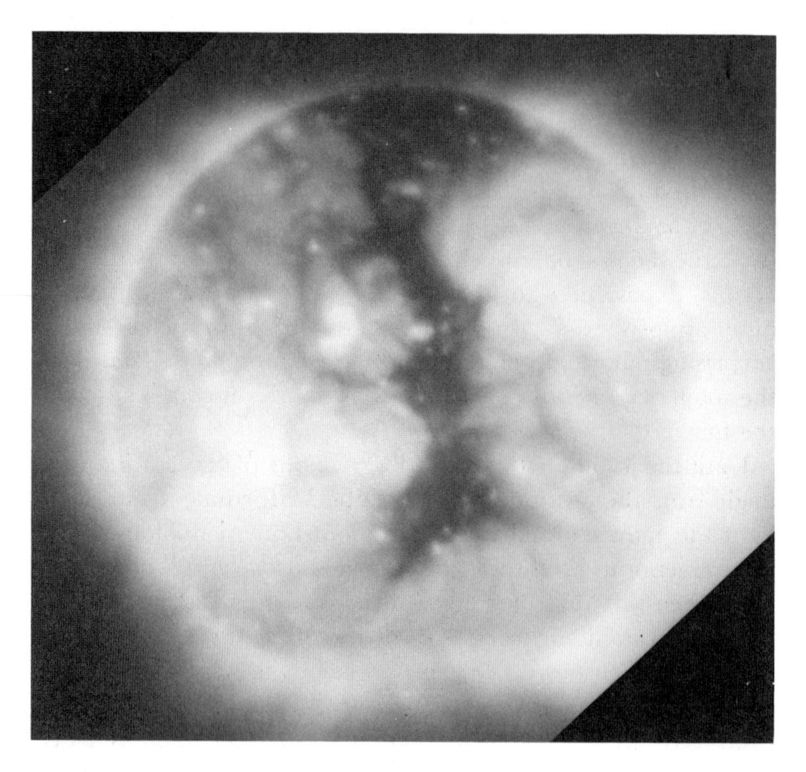

Figure 6
An image of the Sun in x-rays taken by an instrument aboard the Skylab satellite. (Photograph from American Science and Engineering, Harvard College Observatory, and NASA)

Figure 7
Loops on the surface of the Sun. (Photograph from Smithsonian Astrophysical Observatory)

down on it. This is possible because the solar surface, which lies directly beneath, is far too cold to emit appreciable x-rays that could confuse the image of the corona. We see immediately from figure 6 that the corona is highly inhomogeneous. Perhaps this is not surprising in view of the structure visible in visible-light eclipse photographs (figure 1), but the fine details of the x-ray corona far exceed eclipse photographs in their richness.

The varied structure of the x-ray corona is a warning signal to theories of shock wave heating, for the turbulent motions of granulation (figure 5) occur uniformly all over the Sun, except at the location of sunspots. In addition, the shapes of the structures are very highly ordered, apparently by some agent within the corona itself. The brightest structures generally appear in the form of loops, sometimes so closely packed that their individual shapes are hard to discern. When seen in isolation at the limb of the Sun, however, as in figure 7, the graceful symmetry of these loops is readily apparent. The agent that imparts such delicate structure to coronal features is the Sun's magnetic field.

After penetrating the surface from the interior and creating sunspots and surrounding magnetized areas called plages, it then arches through the corona, following the elegant geometry of potential theory, before returning back to the surface at a second sunspot or plage.

The fact that magnetic fields in the corona shape the dense emitting coronal material by constraining it to follow field directions is now understood both theoretically and through laboratory experiments, which show how difficult it is for hot ionized gas, known as plasma, to move perpendicular to the compass direction defined by the magnetic field. The observational proof of this assertion is contained in figure 8, which shows (at left) a magnetogram, or map of solar surface magnetic field structures in which dark areas represent magnetic fields that would attract one end of a compass needle and light areas represent magnetic fields that would attract the other end of the needle (dark and light areas have different polarities). At right in the figure is a (negative) Skylab photograph of the

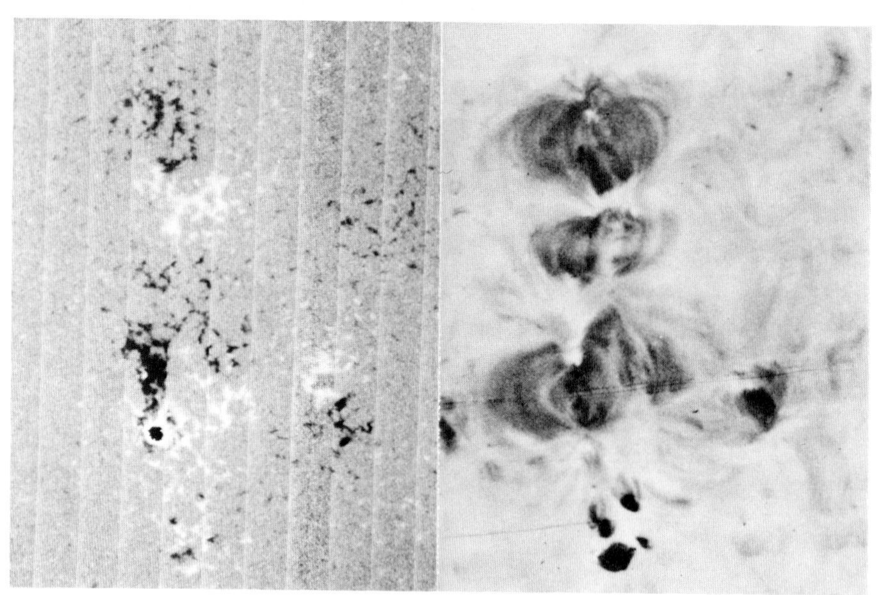

Figure 8
At left, a magnetogram showing solar surface magnetic field structures; and at right, Skylab photograph in ultraviolet of same area. (Magnetogram from Kitt Peak National Observatory; photograph from Naval Research Laboratory and NASA)

corona seen in the extreme ultraviolet light from iron with fourteen of its electrons removed, at a temperature of some 3 million degrees. The hot coronal loop structures (black in this negative image) invariably arch from a footpoint of one polarity up into the corona and return to a footpoint of the opposite polarity. Skeptics about magnetic heating might maintain that although the magnetic field confines the hot corona, something else (like shock waves) could still be heating it within these confined regions. However, recent searches from spacecraft looking for signs of such waves have failed to detect them, and thus the theory of shock-wave heating seems harder and harder to defend. Although the new theory that will evolve to replace it may or may not retain some vestige of the importance of wave motions, it is clear at the very least that the magnetic field plays the fundamental role.

Sunspots, created by strong magnetic fields, were one of two obvious features on the surface that betrayed the existence in the Sun of forms of energy other than heat energy. It may seem like just bad luck that theoreticians thirty years ago picked the wrong one of the two energy sources as a way to solve the puzzle of the hot corona. However, in the 1940s the new field of magnetohydrodynamics, upon which magnetic heating theory rests, was just being born, and by the time it had reached enough maturity to be applied to the problem of coronal heating, shock wave heating theory was already well fixed in people's minds. The lesson perhaps is that one should never underestimate the power of an entrenched idea, even in scientific research.

Returning to figure 6 for a moment, we see a large, dark area of very low coronal emission, stretching practically across the x-ray disk. Such dark regions are aptly named corona holes, for they are great voids in the corona where there is very little coronal material to emit. What little material exists in these regions is considerably less hot than the surrounding corona. One might argue that here is a place where coronal heating really is absent. However, a recent remarkable discovery has shown that high-speed streams of particles spew out from coronal holes; this is known as the solar wind. As the Sun rotates on its axis, these streams of particles shower the interplanetary medium and the Earth, creating Northern Lights and perhaps subtly affecting even our weather.

Coronal holes appear to be places where the magnetic bottle confining the hot high-temperature corona has burst, allowing its contents to spew outward in the solar wind. When astronomers measure the energy contained in the solar wind as it flows out of coronal holes, it seems that it is roughly the same (per unit area) as the energy bottled up in the magnetically confined dense and hot corona. Thus coronal heating appears to occur also in coronal holes, and it is a challenge to new magnetic heating theories to explain how this heating can be almost equally effective both in magnetically closed or bottled-up regions and in the magnetically open coronal holes.

One aspect of coronal holes appears to relate back to the origins of magnetic fields themselves. Solar magnetic fields, as Robert Rosner discusses, owe their existence to the rotation of the Sun. Yet as the solar wind spews out of coronal holes, it takes away angular momentum and over the 5 billion year history of the Sun may have caused it to slow down greatly from its original speed. This in turn suggests that solar surface magnetic fields may be much less strong now than when the Sun was a much younger star, and consequently that the young Sun's corona was denser and hotter. As Rosner discusses, astronomers are beginning to learn by observations of x-ray emission from coronas around other stars to what extent such an inferred history of our Sun's corona makes sense. The key will surely require a far better understanding than we now have of how magnetic fields cause coronal heating and the solar wind. We have made progress in emerging from earlier blind alleys and have some confidence that we are now on the right track, but in many detailed aspects the puzzle of the Sun's hot corona still remains.

ROBERT ROSNER

SOLAR SCENARIOS

The evolution of our understanding of the solar corona has followed a tortuous path and, as Robert Noyes discussed, has been marked by surprisingly tenacious adherence to theories that were not well supported by observations. How did this situation arise? Few (if any) solar astronomers subscribe to Dirac's aphorism that "it is more important to have beauty in one's equations than to have them fit experiments." Although the historical interaction between theory and observation in solar astronomy undoubtedly has molded our subject's intellectual evolution, comparison between theory and observation necessarily has been indirect and furthermore has been influenced crucially by related research focusing on stellar activity and evolution of other stars. Examination of this larger context in which solar coronal theory developed is the aim of this discussion.

The problem of connecting theory to observations is schematically illustrated in figure 1. The connection between the two is indirect because of two distinct reasons: first, the physical system is generally so complex that a rigorous, complete theory that directly accounts for the data cannot be constructed, and second, the laboratory approach to experimental studies, which includes the notion of control experiments, is not feasible for most astrophysical systems. Instead the connection is established by means of a scenario, which we might regard as a kind of plausible story connecting theory with data. In the language of formal logic, the scenario is a metatheory that provides both tools and framework for comparing theory and observations. The simplest scenario for solar activity is one in which convection-generated surface turbulence somehow couples to the corona, thereby heating it; and in this case, because the vigor of convection establishes the level of surface turbulence, convection alone will determine coronal activity levels. For the moment, let us adopt this scenario as a working hypothesis and follow its consequences.

We begin by asking how we might test the validity of this

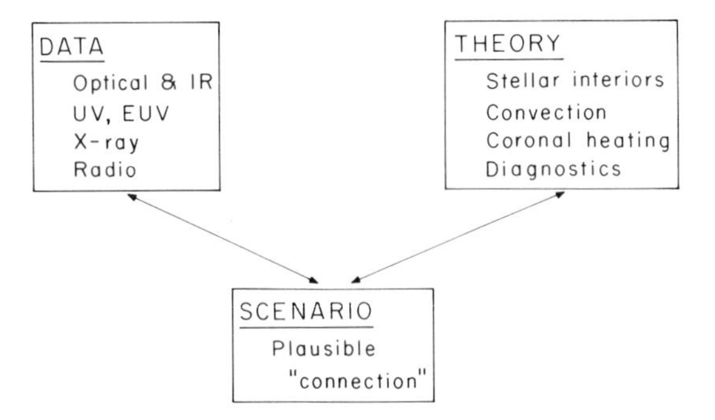

Figure 1
Schematic of the connection between solar observations and solar theory established by the scenario, which provides a nonrigorous framework in which data and theory can be confronted (after M. Rees). The categories listed under "Data" refer to different regions of the electromagnetic spectrum; IR, UV, and EUV stand for infrared, ultraviolet, and extreme ultraviolet, respectively.

scenario. The first step involves development of detailed theories. The scenario itself then provides an interpretive framework in which theoretical consequences are developed as observables and ultimately compared with observations. If the theoretical constructions are aimed at solar observations alone, then we must follow the traditional path of solar physics, as Robert Noyes outlined, which leads to the realization that magnetic fields are crucial to coronal heating. Of course, it can be argued that the physics of the scenario must be common to that class of stars similar to the Sun, and if we regard the Sun as a prototype, our scenario will lead to theoretical predictions about the behavior of other stars. This is the path I intend to pursue. Remarkably, in the past this same solar-stellar analogy provided the strongest support for our assumed coronal scenario, despite the fact that no direct observations of stellar coronas were available until the mid-1970s.

The internal structure of a Sun-like star is schematically represented in figure 2. Its nuclear-burning core provides the energy to maintain the star as a whole, and in the star's interior, energy is carried outward primarily by radiation. It is now thought that if a star's surface temperature lies below ~7500 K (or equivalently if its mass is less than approximately 2 solar masses), the outer envelope of the star becomes too opaque for effective energy transport by radiation. Instead fluid motions—convection—take over this function in much the way convection

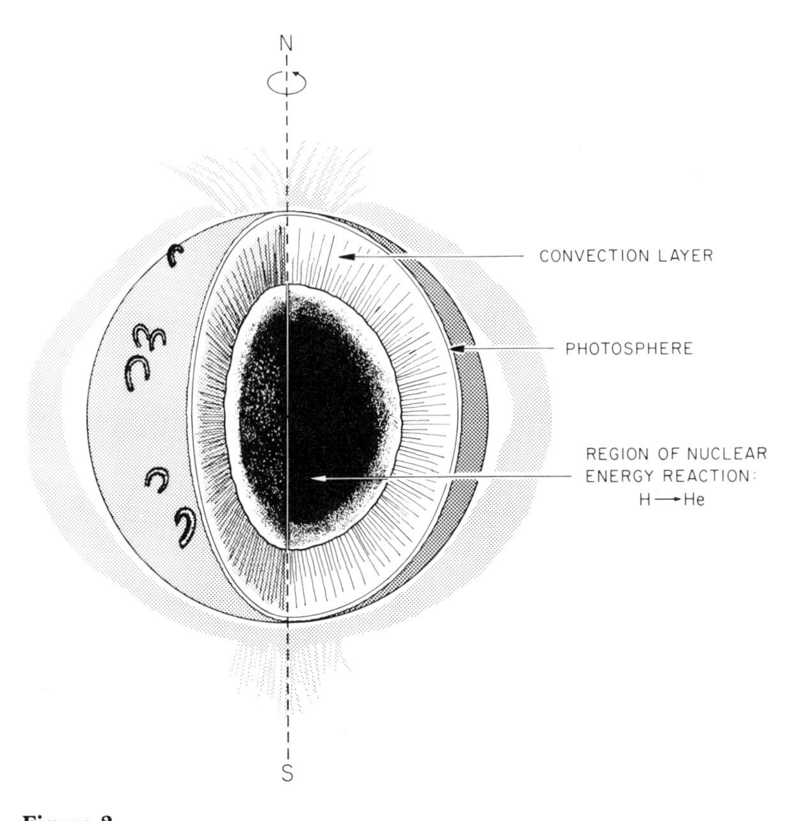

Figure 2
Conceptual drawing of the Sun's structure, showing the nuclear-burning core, in which the energy powering the Sun is currently released and carried outward by photons; the convection zone, which forms an outer layer of circulating gas in which energy is carried outward more efficiently by gas motions than by photons; and the photosphere, which forms the visible surface of the Sun and is itself subjected to vigorous pummelling by the gas motions of the underlying convection zone. Overlying the photosphere is the tenuous outer solar atmosphere, whose high temperature relative to that of the photosphere is still poorly understood.

supplants thermal conduction as the primary energy-transport process in a pot of boiling water. Thus all stars roughly less massive than 2 solar masses are expected to have a vigorously convecting surface envelope. The precise extent (depth) of this envelope and the vigor of the associated fluid motions depend on an individual star's mass and composition.

Detailed theories of stellar interiors and of turbulent fluids predict, given a star's mass, the level of noise generated by convective fluid motions once they break through the star's surface. Figure 3 exemplifies the results of such a calculation. It shows how the noise level emitted per square centimeter at the star's surface varies with the star's mass. For our purposes, three facts stand out. First, the Sun's surface noise level apparently lies near the peak of the curve. If we suppose a direct

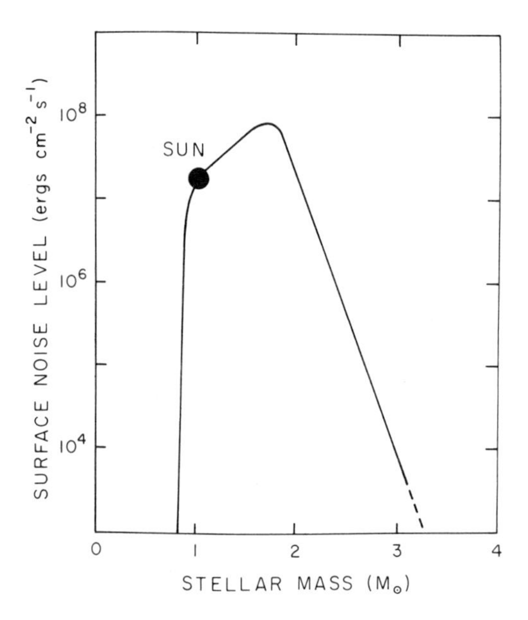

Figure 3
Theoretical calculations of the energy radiated outward by turbulent gas motions (in the form of sound waves) have been carried out for stars of different masses. The graph shows a typical result of such a calculation; the amount of noise energy emitted radially outward per unit area per second for stars of different masses (measured in terms of the Sun's mass). The energy is measured in units of ergs; the energy released on impact of a penny dropped from waist high is about 1 million (10^6) ergs. The amount of energy supplied by turbulent motions establishes an upper limit to the amount of energy that the hot, outer, tenuous atmosphere (chromosphere and corona) can radiate.

connection between convective noise and coronal heating (as would be the case if we hypothesize an acoustically heated corona), we would expect the Sun's corona to be correspondingly bright. Second, the surface noise level plummets as stars of mass either somewhat larger or smaller than the Sun's are considered. Given our initial scenario, these stars should have far less bright coronas than the Sun. (Our galaxy's star population is very heavily weighted toward stars of low mass; in fact, stars whose mass is less than half the Sun's mass are thought to be over ten times more frequent than stars of mass similar to that of the Sun. The Sun and the other stars with high surface noise levels are relatively rare. Most stars encountered in the galaxy should therefore be far weaker coronal emitters than the Sun.) Third, the theory predicts a unique value for the surface noise level for every stellar mass, in agreement with the scenario's insistence that convection alone determines the coronal emission level. Our initial scenario therefore predicts that the Sun's corona ought to be typical of stars of similar mass, that the Sun's coronal emission lies at the upper range of stellar coronal emission levels, and that solar-like emission levels are rare (for example, the numerous low-mass stars ought to be weak coronal emitters when compared with the Sun).

Before the mid-1970s there were no direct observations of coronal emission from solar-like stars. As Robert Noyes points out, the temperature of stellar coronas is so high that the radiation is predominantly ultraviolet and x-ray rather than visible light. But even satellite-borne instruments capable of circumventing the absorption of ultraviolet and x-ray photons by the terrestrial atmosphere were not sensitive enough to detect the relatively weak stellar emission. Instead the burden of proof for supporting our scenario was placed on two indirect coronal indicators.

The first indicator took advantage of the fact that on the Sun there are good correlations between coronal activity levels and surface (photospheric) magnetic activity and between surface magnetic activity and radiation by singly ionized calcium atoms (Ca II) in the Sun's surface layers. On the Sun, regions of bright Ca II emission (the plages) coincide with the regions of active surface magnetic fields and vigorous coronal activity. Furthermore Ca II emission has long been observed in the spectrum of solar-like stars thought to have surface convection zones, and it seemed natural to attribute this emission to plage regions similar

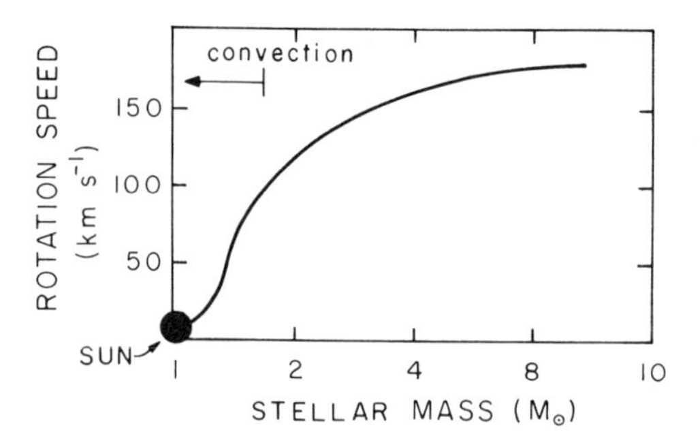

Figure 4
Observations of the rotation rate of stars near the Sun allow astronomers to study the variation of the surface rotation speed (for example, the speed at which the star's equator moves) with stellar mass. The curve shown here demonstrates the average behavior of the equatorial rotation speed (in kilometers per second) of stars of different mass (as measured in units of the Sun's mass).

to solar active regions on the star's surface. Thus emission by singly ionized calcium has been taken as a tracer for stellar coronal activity and has been extensively studied, following the pioneering work of O. C. Wilson.

The second indirect probe for coronal activity is the stellar rotation rate. If one measures the average surface rotation speed of stars and studies its variation with stellar mass, the curve shown in figure 4 is obtained. The pertinent feature of this curve is that the point where the surface speed of a star rapidly declines with decreasing mass coincides with the stellar mass at which surface convection zones are thought to begin their existence. This coincidence is not thought to be accidental; our own scenario claims that stars with convection zones should have coronas. Coronas are associated with mass loss, in analogy to the Sun's solar wind, and such mass loss leads to a slowdown in stellar rotation, much as a skater's spin is reduced by an extension of the arms. Thus if all stars begin their life with comparable initial conditions, those with convection zones should show a systematic slowdown with age, and those without convection zones should show no such age effect. In a now-classic study, R. Kraft showed that this effect is precisely observed and that it provides a natural explanation for the data shown in figure 4.

What can we conclude? In spite of the absence of direct coronal observations, the data I have discussed give fairly convincing support to my simple scenario. In particular, the scenario provides a very plausible explanation for a previously incomprehensible set of data—the variation of stellar rotation rate with mass and time—by tying these data to the presence of convection and coronal plasma, as predicted by detailed theory and supported, indirectly, by stellar Ca II emission. It thus unifies a number of previously disparate sets of observations under a common, relatively simple metatheoretical umbrella. This is the usual mark of a successful theory, and it is not surprising that the scenario found wide acceptance. Its acceptance was not hindered by the fact that it fit in well with then-current conceptions of solar coronal heating theory. My simple scenario thus buttressed a solar coronal theory—acoustic coronal heating—which in the solar context had no direct observational support.

But what about direct stellar coronal observations? Relatively little was available until the end of the 1970s, when data from two satellites, the International Ultraviolet Explorer and the Einstein X-Ray Observatory, began to provide a wealth of information on the coronas of other stars. The most dramatic impact has been made by pictures acquired from the latter instrument in an extensive stellar coronal survey by G. S. Vaiana in collaboration with a large group of American and European scientists. I will focus on these x-ray results, which give a direct measure of stellar coronal activity. Indeed stronger x-ray emission should indicate high coronal activity.

The Einstein Observatory is an orbiting satellite that carries a telescope capable of forming images of x-ray emitting objects much in the way that optical telescopes form images of objects in visible light. The technique of x-ray observations has a long ancestry. For example, Robert Noyes has discussed an image of the Sun taken photographically at x-ray wavelengths by a telescope onboard the now-defunct Skylab space station. (See Noyes's figure 6.) This solar x-ray telescope was the direct predecessor of the Einstein Observatory instrument, having been conceived and designed by the same group of scientists led by R. Giacconi.

Stars are, of course, much too far away to be resolved spatially, as the Sun has been in figure 6 of Noyes. Figure 5 is an x-ray image of the star π^1 Ursa Majoris. The electronically recorded picture of the star's x-ray emission can be seen in the center of the image. Knowing its distance from us and measuring its

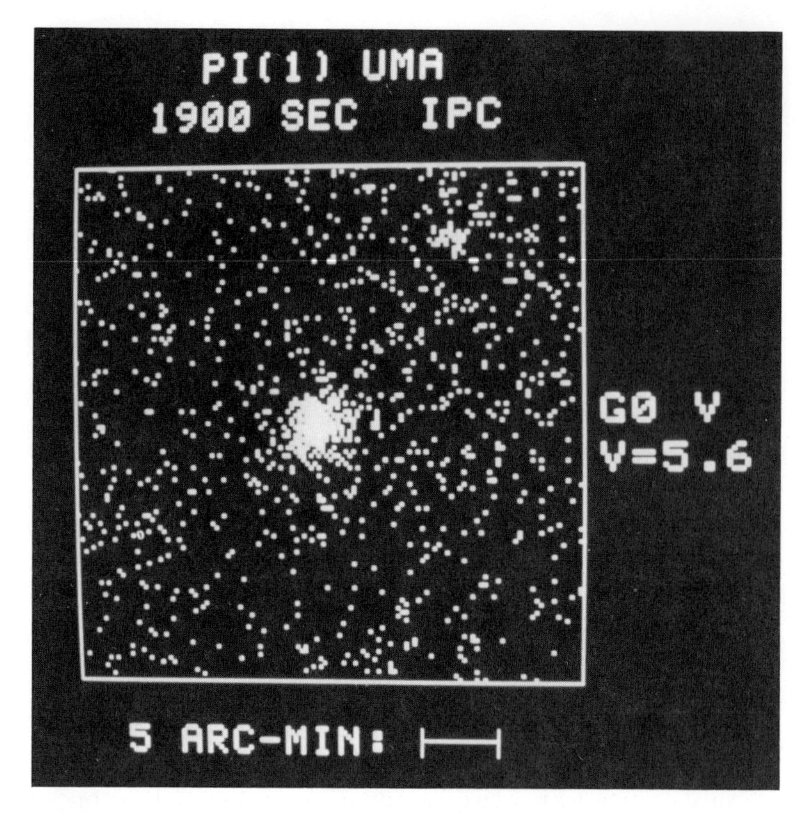

Figure 5
A solar-like star, π^1 Ursa Majoris, as seen in x-rays. This image was electronically recorded by a photon-counting detector in 1979 on board the Einstein Observatory, by use of a large imaging x-ray telescope. The many individual, uniformly distributed spots represent noise; the star itself is seen near the center of the field of view as an enhancement in the number of spots. (Photograph by the courtesy of Riccardo Giacconi)

apparent brightness allows us to deduce the intrinsic brightness of π^1 Ursa Majoris's corona. Although its mass is roughly equal to that of the Sun, it is clearly far brighter in x-rays; in fact, its x-ray output is over one hundred times larger than the Sun's coronal emission level. This star is thought to be comparatively young and shows many signs of stellar youth; it is a rapid rotator and is bright in Ca II emission. Evidently stellar mass is not the sole determinant of coronal activity; yet according to our scenario, stellar age should not figure in fixing coronal emission levels. As far as stellar x-ray astronomers are concerned, the contradiction between observation and theoretical prediction is fortunate; it opens the possibility that the Sun is atypical and that the x-ray sky is more interesting than our scenario predicted.

According to our initial scenario stars of lower mass than the Sun should have lower coronal activity levels. An easy way to check this assertion is to look at a multiple star system that contains at least one star similar to the Sun, as well as stars of radically different mass. The closest star system to us, α Centauri, contains a virtual twin to the Sun, α Centauri A, as well as a less massive star (named α Centauri B). Figure 6 shows an Einstein Observatory x-ray image of this star system. Remarkably it is the less massive star that is the brighter x-ray emitter, completely contradicting our scenario prediction. This observation excludes a possible relative age effect on coronal emission because these two stars are of equal age. Effects due solely to the binary nature of these stars are also excluded by the large separation between the two components.

What have we learned from stellar observations? We now know that all solar-like stars are x-ray emitters and, by implication, have coronas; that the coronal activity level does not correlate well with stellar mass and therefore seems to be insensitive to the vigor of surface convection; and that the Sun's corona is not at all representative even of stars with comparable mass (in fact, its coronal emission level is very weak when compared to younger stars of similar mass). We also are forced to conclude that the basic assumption of our scenario—that convection alone determines coronal emission—must be wrong. This is a remarkable result because our simple scenario seemed so convincing. Moreover it is not obvious how the successes of the simple scenario can be incorporated (as they must) in a new scenario that accounts for the stellar x-ray observations as well.

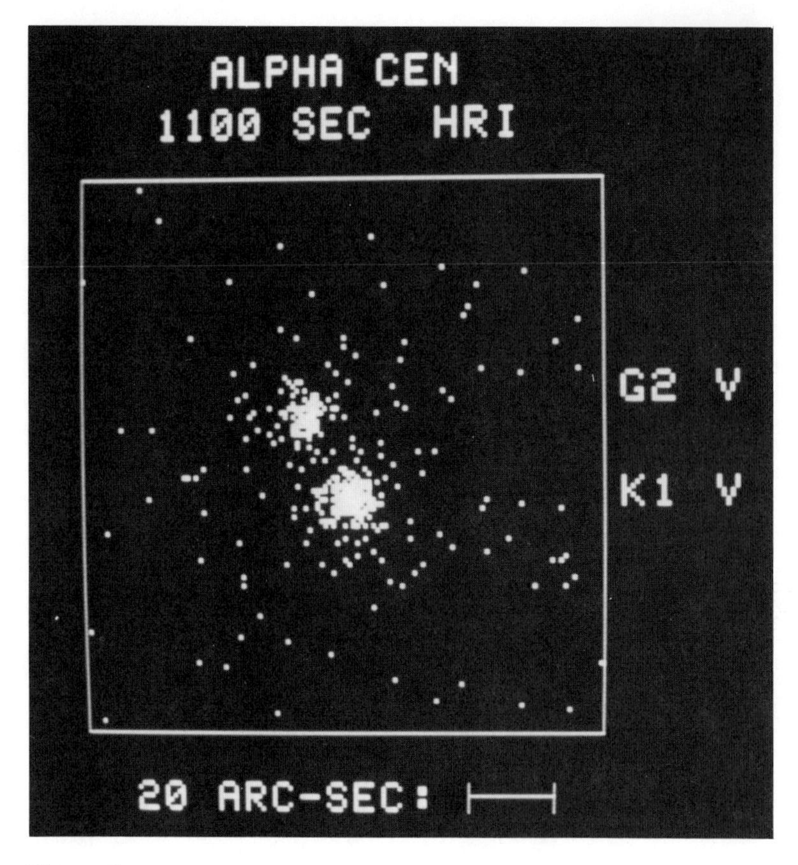

Figure 6
The nearest binary star system, α Centauri, provides a vivid example of the difficulties encountered by classical coronal theory. As shown in this Einstein Observatory x-ray image, the less massive star, α Centauri B, is the brighter star in x-rays, contrary to expectations based on classical coronal theory. The latter predicts (see figure 3) that the less massive star should have a far less vigorous corona. (Photograph by the courtesy of Riccardo Giacconi)

In order to proceed, we must resign ourselves to the fact that stellar mass alone is not the determinant of coronal activity and that additional intrinsic physical parameters must figure in a revised coronal scenario. We now have preliminary indications that one previously missing important variable is the stellar rotation rate. Very recent observational evidence suggests that rapid rotation is correlated with relatively enhanced coronal activity levels. If this is the case, we can now develop a revised scenario that encompasses what we have learned in both the solar and stellar contexts.

Consider a rotating star with an outer convection zone (figure 7). The coupling of rotation and convection can be shown to amplify magnetic fields in the star's outer layer; this process of magnetic-field production is called a magnetic dynamo. The magnetic fields produced by convection and rotation are buoyant and emerge from the interior but continue to be jostled by turbulent fluid motions at the star's surface. This jostling or shaking of the magnetic fields distorts them and, in the low-density atmosphere overlying the star's surface, leads to gas heating as the emerged magnetic fields attempt to relax to an undistorted configuration. A hot corona confined by the emerged magnetic fields is produced. The emission from these structures is what we see in the x-ray images. As these fields evolve, the emerged loops increase in size and eventually break open. The result is a gaseous wind, and thus a loss of mass from the star. This mass loss feeds back to the star's rotation by slowing down the star; it therefore decreases the dynamo's effectiveness. We thus find that the coronal phenomenon is truly self-defeating; once initiated, it inevitably leads to its own demise.

The obvious question is whether this revised scenario satisfies the observational constraints I have discussed. The basic connection assumed between magnetic fields and coronal heating does account for the observed correlation among surface magnetic fields, Ca II emission, and coronal activity. The virtual independence of coronal activity levels from surface convective turbulence levels is not so easily explained. Although magnetic-field-related coronal heating is expected to be less sensitive to the level of convective surface activity than acoustic coronal heating, a quantitative theory is lacking. However, both the variation of coronal activity during the solar cycle as well as the detailed correlation of solar coronal emission with the surface

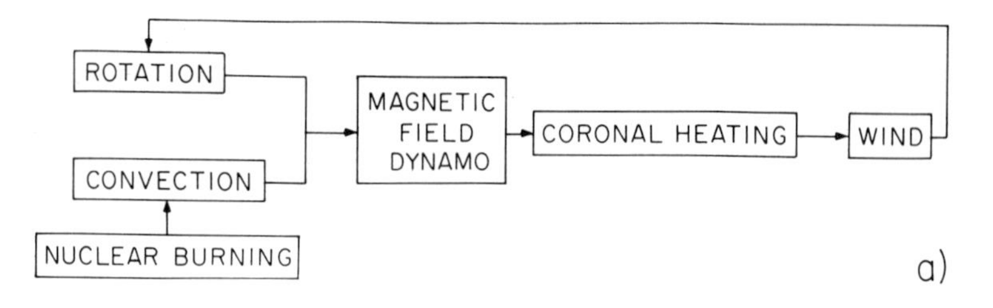

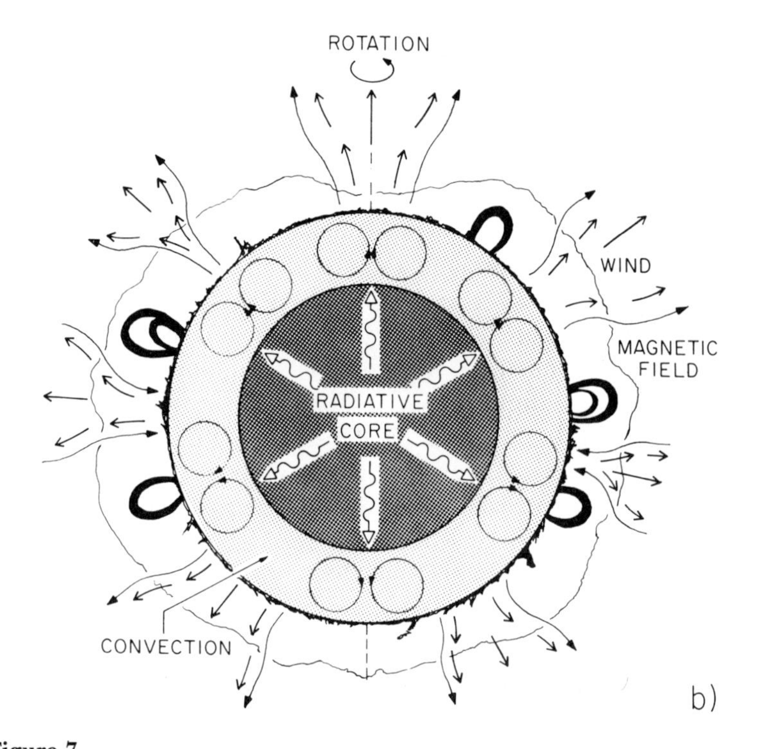

Figure 7
Schematic of a possible revised coronal scenario. The flowchart (*a*) shows the dynamical connections between the various processes that ultimately lead to x-ray emission from a hot corona, as graphically illustrated in *b*. The ultimate energy source is, of course, energy release in the nuclear-burning stellar core.

magnetic field suggest that the coronal activity level is primarily determined by the amount of magnetic flux (magnetic field strength multiplied by area covered by magnetic flux tubes) present; why this is so is as yet unknown. It is important to realize that the various theoretical ingredients in the new scenario (figure 7) are themselves not new. What has happened is that the plausible, but nonrigorous, story connecting theory with observations has had to be revised significantly as new data were obtained.

What have we learned about the Sun by regarding it as a prototypical star? From the theoretical point of view, we have discovered a significant new variable, stellar rotation, which figures in determining the coronal activity level. This discovery is consistent with the new scenario based on the notion of a magnetically heated corona developed in the purely solar context. Most crucially, it could not have been obtained by looking only at the Sun. (The Sun, after all, is associated with one rotation rate, so that effects due to variation in rotation rate cannot be studied.) Similarly we conclude that when the Sun was young, it must have been far more active than it is today. Because coronal emission occurs at wavelengths of radiation to which chemical reactions, particularly organic reactions, are quite sensitive, we may have to rethink the processes that led to today's atmosphere and consider the mutational effects of the radiation on early life forms. Both of these lessons are encompassed within the new scenario. But recall our original caveat: a scenario is nothing but a plausible story connecting theory and data. As we learn more about the Sun and other stars like it, we can hope to increase the plausibility of the scenario, but we should expect to be surprised and to change the scenario once again, just as we have in the past.

Further Reading

Dupree, Andrea, ed. "Cool Stars, Stellar Systems and the Sun." *SAO Special Report No. 389*. Cambridge: Smithsonian Astrophysical Observatory, 1980.

Eddy, John A. *The New Sun: The Solar Results from Skylab*. Washington: NASA, 1979.

Gibson, Edward G. *The Quiet Sun*. Washington: NASA 1973.

Kraft, R. P. "Studies of Stellar Rotation: V. The Dependence of Rotation on Age Among Solar-Type Stars." *The Astrophysical Journal* 160 (1967): 155.

Noyes, Robert W. "New Developments in Solar Research." In *Frontiers of Astrophysics,* edited by E. H. Avrett. Cambridge: Harvard University Press, 1976.

———. *The Sun.* Cambridge: Harvard University Press, 1981.

Parker, E. N. "The Solar Wind." *Scientific American* (April 1964).

Pasachoff, J. M. "The Solar Corona." *Scientific American* (October 1973).

Vaiana, G. S., and Rosner, R. "Recent Advances in Coronal Physics." *Annual Review of Astronomy and Astrophysics* 16 (1978): 393.

Wilson, O. C. "Stellar Chromospheres." *Science* 151 (1966): 1497.

Withbroe, G. L., and Noyes, R. W. "Mass and Energy Flow in the Solar Chromosphere and Corona." *Annual Review of Astronomy and Astrophysics* 15 (1977): 393.

CHAPTER 5

COSMIC POWERHOUSES: QUASARS AND BLACK HOLES

HARVEY TANANBAUM

THE FARTHEST AND THE BRIGHTEST

Our ability to study the universe has been dependent primarily on information received in visible light. First with the unaided eye and later with telescopes, we have studied the Sun, Moon, planets, stars, and galaxies. From detailed observations in visible light, we have obtained measurements of distances, masses, temperatures, total energy outputs, and ages for many astronomical objects. These data have formed the basis for a standard astronomical view of the universe.

This approach to astronomy began to change with the realization that some stars and galaxies can emit a substantial fraction of their energy in forms other than visible light. The nearest star, our Sun, has now been observed to radiate radio signals, infrared rays, ultraviolet rays, x-rays, and gamma rays. These radiations are all forms of electromagnetic energy; yet for the Sun, most of the energy output still is in visible light.

Some other objects radiate a much larger fraction of their energy in radio signals or in x-rays. The study of such objects can provide us with new perspectives of the violent processes that often take place in nature. The discovery and study of quasars provide a particularly fascinating example and is the basis for this discussion and the one by Alan Lightman.

In the 1950s a number of astronomical radio sources had been detected with sensitive radio receivers or telescopes. Some of these sources were identified with very faint, distant galaxies, and some were not identified with any particular visible light counterpart. Then at the December 1960 meeting of the American Astronomical Society, a group of astronomers (T. A. Matthews, J. G. Bolton, J. L. Greenstein, G. Munch, and A. R. Sandage) made the first of what would be a series of startling announcements. One of the radio sources, Number 48 in the *Third Cambridge Catalogue* of radio sources and therefore called 3C48, had been pinpointed on the sky, and at that location a relatively faint star—not a galaxy—was found. Over the next two years several more stars were located at the positions of unidentified radio sources. Studies of the light content of

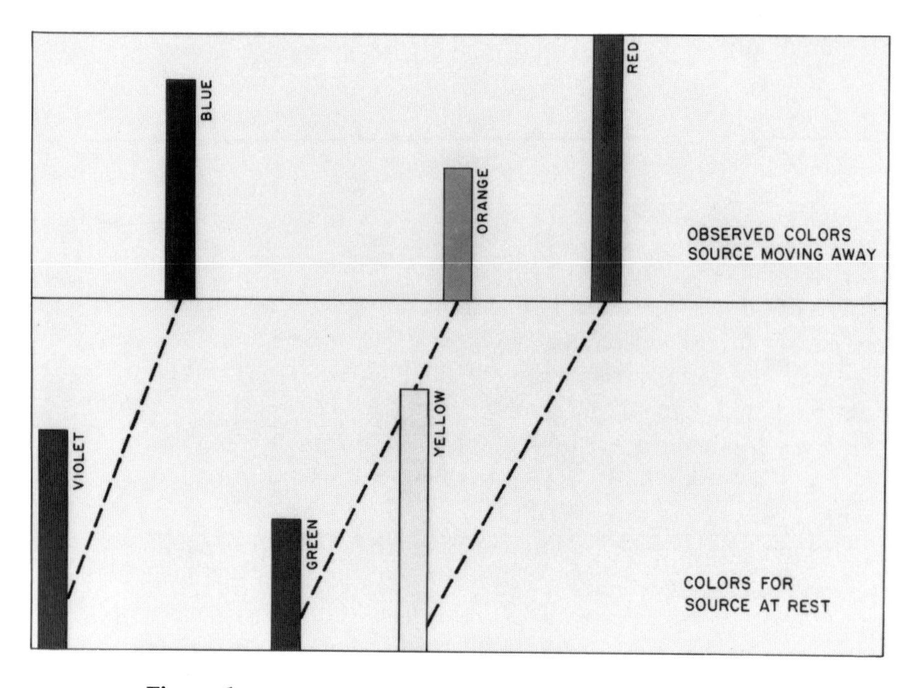

Figure 1
Simplified representation of light color pattern observed for source moving away (top) and corresponding light color pattern for source at rest (bottom).

these stars proved even more perplexing for they did not look like any known stars of galaxies either.

Figure 1 presents a very simplified description of this situation, representing the observations and their eventual interpretation. The top half of the figure shows how one of these radio stars might have appeared if its light were separated into its various wavelengths or colors. (With a prism white light can be separated into a spectrum of colors—red, orange, yellow, green, blue, indigo, and violet.) If a series of filters or a prism were used with the telescope, we might find the light from one of these radio stars separated into the shown patterns of colors, with the red light brightest, the blue light almost as bright, and, in between these two extremes, some fainter orange light. If this pattern did not correspond to anything seen before, however, we could not understand the nature of the star. This was the case for the stars identified with the radio sources at the beginning of the 1960s. In 1962, however, the astronomer Maarten Schmidt discovered that the light pattern for a star identified with the radio source 3C273 could be related to a pattern that was already known. This pattern is represented in the bottom half of figure 1, where the yellow light is most intense, violet is almost as strong, and in the middle there is a fainter trace of green. What Schmidt recognized for 3C273 was that the observed color pattern was the result of a precise shifting of the more standard color pattern. In the context of figure 1, the violet that might have been expected was shifted to the blue, the green to the orange, and the yellow to the red. In short, the observed spectral pattern from these strange starlike sources had not been understood because it corresponded to a large shifting of the expected pattern toward the red. Schmidt's discovery was the major intellectual breakthrough required to show the unusual nature of quasars, as these quasi-stellar radio sources came to be called.

Figure 1 is a greatly simplified description of the actual situation. Schmidt actually used a spectrograph at the focus of the 200-inch Mount Palomar telescope to spread or disperse the visible light from the star identified with 3C273. The dispersed light showed a number of broad emission features (or bumps of excess light) superimposed on a continuum. Schmidt recognized that these emission features corresponded to emission lines produced by electron transitions in hydrogen, oxygen, and magnesium atoms, but the wavelengths of the observed lines differed from those normally observed by a factor of 1.158.

This redshifting of the color in light is an effect well known in physics and astronomy and occurs when the light source is moving away from the observer. The effect is called the Doppler shift, after the scientist who first described it. The Doppler shift also occurs with sound waves; the classic example is illustrated in figure 2. As a train approaches an observer, the sound of its whistle is high pitched; as the train passes, the sound is a little lower pitched; and as the train moves away the whistle becomes still lower pitched. Moreover, the greater the velocity of the train, the greater will be the shift in pitch of its whistle. This effect can be observed by listening carefully for sound changes when an ambulance or a police car with a siren passes by. The same effect was described by Doppler for visible light and is shown schematically in figure 2 by the moving light bulb. An approaching light source will have its normally yellow at-rest color shifted to the blue, while a light source moving away will have its color shifted to the red. To produce a color change that

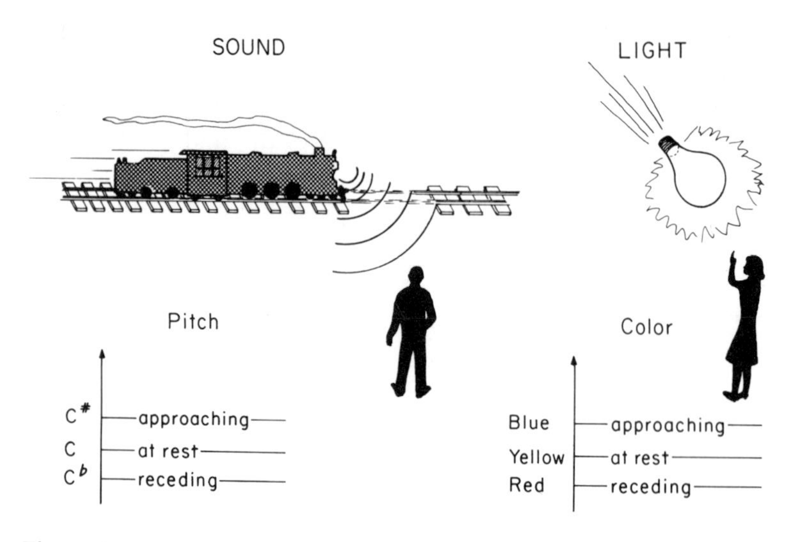

Figure 2
Description of the Doppler shift, showing the change in sound pitch for a moving train and in light color for moving light source.

can be observed with the naked eye, the light source must be moving at a measurable fraction of the speed of light, which is much greater than the speed of sound. Since objects we normally encounter are not traveling at a sufficiently high speed, the Doppler shift in light color is not observed in our everyday lives.

For the colors to be shifted by the amount noted by Maarten Schmidt, 3C273 must be moving at about 15 percent of the speed of light. Since the colors are red shifted, this means 3C273 is moving away from us at this very high speed. Furthermore, as long ago as 1929, Edwin Hubble had used his studies of galaxies to show that the faster a galaxy was moving away from us, the farther away it was. This makes sense in the view of an expanding universe. Thus 3C273 and other quasars that were identified afterward are believed to be at very great distances (since their red shifts all indicate that they are moving away very rapidly). In fact most of the quasars are farther away than any observed galaxies and are the most distant objects known in the universe. By using the red shift to calculate the distance and by taking into account the observed brightness of a quasar, we can compute the total energy being radiated. This energy output is very large, often one hundred to one thousand times more than a whole galaxy typically radiates. It indicates that in spite of their starlike appearance, the quasars are radiating far too much energy to be stars. This poses one of the major questions to be answered experimentally and theoretically: How does a quasar produce so much energy?

Fortunately, we now have an additional tool both to discover and to study quasars: the measurement of their x-ray properties with a special x-ray telescope carried on board a space satellite named Einstein, in honor of the famous scientist. Figure 3 shows an x-ray picture taken of the radio quasar 3C47. The bright source in the middle of the figure is that quasar, seen in its x-ray light. A second bright source about one-quarter as intense as the quasar is also seen in the x-ray picture. (The other little dots are not sources but background caused by radiation moving randomly in space.) We have now observed more than two hundred quasars with the Einstein satellite, and almost all of them are bright x-ray emitters. Furthermore, work on quasars over the past twenty years has shown that less than 10 percent of them are detectable as radio sources, and many of them, especially those most distant, cannot be distinguished from stars at first study in visible light. This means that the x-ray emitting property may be the most universal feature that can be used to select or find quasars.

Why are we concerned with having a simple means of selecting quasars? Why can we not look at their detailed visible light spectra or color distribution and measure their red shifts as Maarten Schmidt did? The answer may be clearer after looking at the visible light counterparts of these two x-ray sources.

Figure 4 is a photograph taken with a large visible light (optical) telescope. The orientation and scale are the same as in the x-ray picture shown in figure 3. The arrows indicate the visible light counterparts to the quasar 3C47 and to the second x-ray source. This second source is identified with a fairly bright star, which is too faint to be seen with the unaided eye but which can be seen with a good set of binoculars. The optical counterpart of the quasar is five thousand times fainter than the star. It requires a large telescope even to be seen, and a detailed study of its light content takes one to two hours of observing time. Furthermore such spectral studies are often done only one object at a time, and there are approximately ten thousand stars or starlike objects in just this small amount of sky that are at least as bright in visible light as 3C47. It would take an optical astronomer more than ten years to obtain spectra for all of these

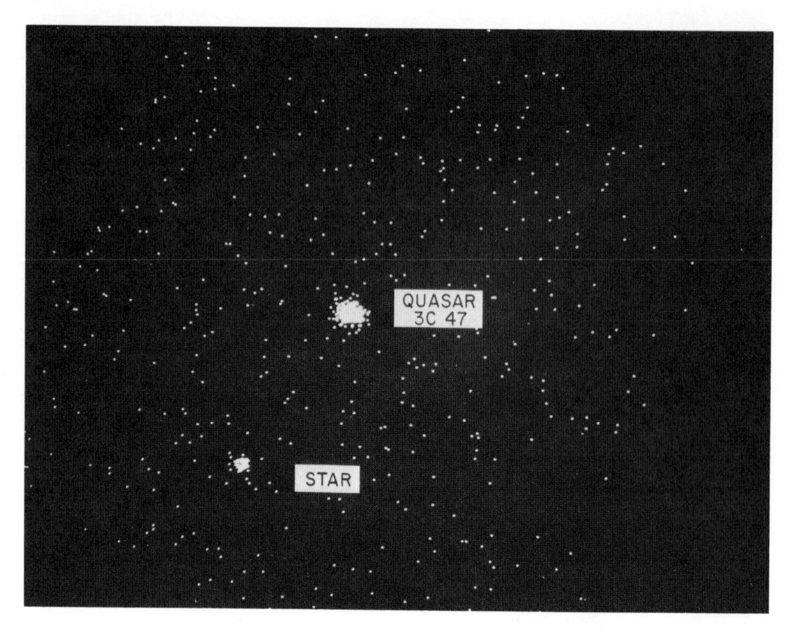

Figure 3
X-ray picture obtained for quasar 3C47 with the Einstein satellite. The quasar is the bright object in the center. A second x-ray source identified with a relatively bright star is also seen. (Photograph from Harvard-Smithsonian Center for Astrophysics)

objects. This is too long, particularly since the radio emission or x-ray emission can be used to point out the quasar very quickly. How do we tell the quasar from the star in the x-ray picture? The amount of energy seen in x-rays compared to that seen in visible light is about twenty thousand times higher for the quasar than for the star in this field. We can use this large excess of x-ray to visible light to distinguish the quasars from the stars.

We are most interested in learning about the energy source in quasars. Some of the quasars have been found to vary in intensity or brightness—in visible light, in radio, and now in x-rays—over times of days and years. Figure 5 shows the x-ray intensity measured for 3C273 for most of three days in December 1978 with the Einstein satellite. Each cross represents the average intensity over one hour; the higher the cross in the figure, the brighter the x-ray emission from 3C273. Time runs from left (the beginning of the observation) to right (the end of the observation). For the first day the x-ray intensity is fairly steady. It is also fairly steady on the second day but at a level 10

Figure 4
Visible light picture (same scale and orientation as in figure 3) showing visible light counterparts for 3C47 and x-ray emitting star, at the tips of the two arrows. The star is approximately five thousand times brighter than the quasar in visible light. (Photograph copyright © National Geographic Society-Palomar Observatory Sky Survey, reproduced by permission)

percent higher than on the first day. Thus these data show that the x-ray intensity increased by about 10 percent in no more than the twelve-hour separation between the two viewing intervals. At the distance of 3C273 this 10 percent change in x-ray brightness corresponds to switching on about 1 trillion Suns in less than one-half day. The time scale of the variability has been used to estimate the size of the emitting region (see the discussion by Lightman), and the region must be very small compared to sizes usually associated with galaxies. On the other hand, the tremendous amount of energy involved and comparisons with certain special galaxies almost guarantee that quasars are associated with the very centers, or nuclei, of galaxies. The production of so much energy in so small a region sets limits on the amount of matter involved in generating the radiation and also requires a relatively high efficiency in the production of the radiation.

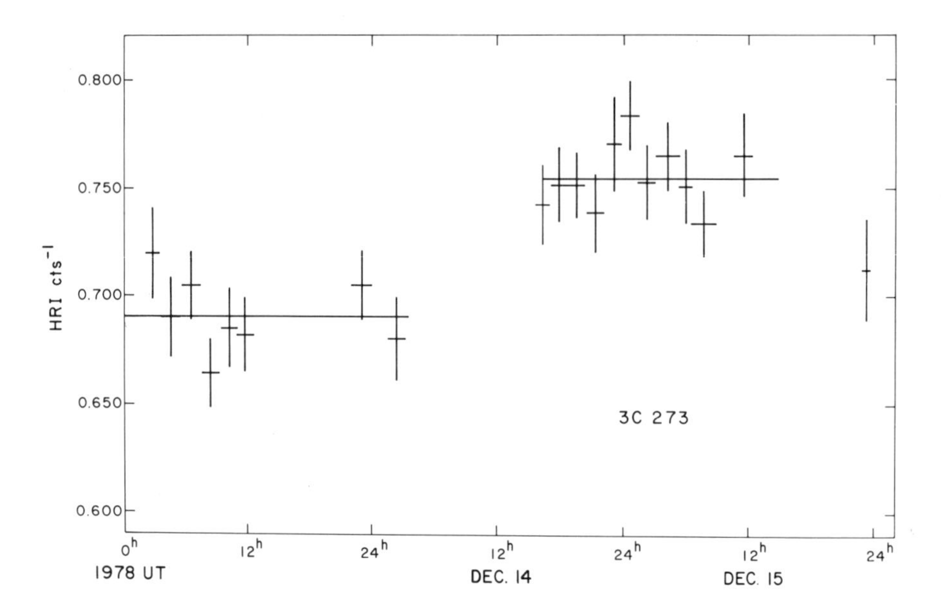

Figure 5
X-ray intensity versus time for quasar 3C273. The intensity obtained in one-hour samples is indicated by crosses, with the height of the cross indicating relative x-ray brightness. Solid lines show average intensity for first day and second day. Time runs from left (December 13, 1978) to right (December 15, 1978). Data show that the intensity increased by approximately 10 percent between three hours and fifteen hours on December 14. (Data courtesy of P. Henry, Harvard-Smithsonian Center for Astrophysics)

Another means of studying the energy generated by quasars is to look at the overall electromagnetic spectrum—from radio signals to gamma rays. Figure 6 shows the intensity of the radiations observed from 3C273 in various energy bands (or colors). Each band is labeled. The lowest energy shown is the radio; in the middle are the infrared, visible light, and ultraviolet; the x-ray and the gamma ray are the highest energy. The overall shape in the figure can be compared to shapes observed for stars and galaxies and can be used to rule out as the energy source a very hot, glowing gas as would be typical for a star. A process called synchrotron radiation may be involved in the radio and possibly the infrared emission. In this process very energetic electrons follow curved paths as they move in a strong magnetic field, and as they move, they produce the observed radiation. In turn the radio and infrared photons (or light particles) can scatter or bounce off the energetic electrons and

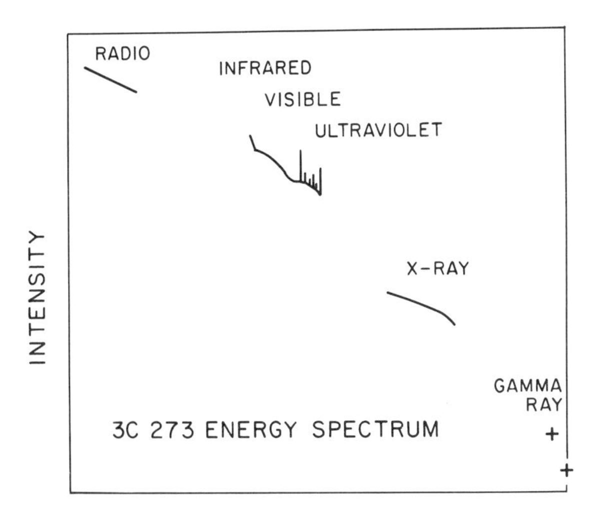

Figure 6
Electromagnetic spectrum for 3C273 showing relative intensity in various energy bands (radio, infrared, visible, ultraviolet, x-ray, and gamma ray).

pick up energy in what is called a Compton scattering process. This mechanism may lead to the x-rays and gamma rays observed, although it does not directly explain how the energetic electrons or strong magnetic field are ultimately powered. Much of this research is not fully developed, but we should be able to learn details of the radiation process from systematic studies of the overall radiation spectra for many different quasars and from monitoring changes with time in various parts of the spectra of individual quasars.

There is an additional observational technique from which we obtain insight on the ultimate energy source in quasars. Figure 7 is a radio picture of the region around the quasar 3C47 shown in figures 3 and 4 in x-ray and visible light. Here the distance scale is greatly expanded so that the x-ray emitting star is far outside the boundaries of this picture. The radio map indicates relative intensity by the series of circular shapes; the closer together the circles, the more intense the radio source. Three sources are apparent in this radio map. One radio source is located in the center, at the position of the optical and x-ray emission from 3C47. Then on each side of the central source is a separate radio source. Each of these sources is about 3 million light-years from the central source. If the particles responsible

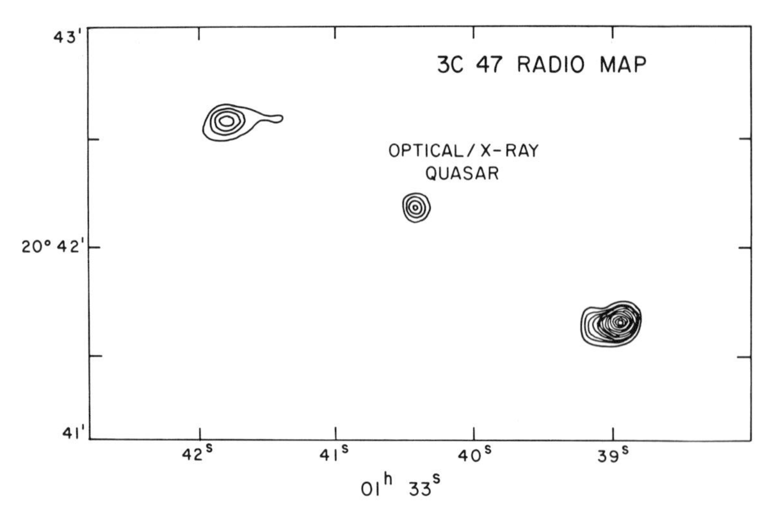

Figure 7
Radio picture of the region around quasar 3C47. The central source is located at the position of optical and x-ray emission. The outer radio sources (lobes) are discussed in text. Radio intensity is indicated by concentric circles, with greater intensity indicated by closer spacing. (Figure adapted from G. G. Pooley and S. N. Henbest, *Monthly Notices Royal Astronomical Society* 169 [1974]: 477)

for the outside radio sources are ejected from the central quasar and travel at the speed of light, they require 3 million years to reach the location shown in the map. Thus the quasar has probably been active for at least 3 million years. This structure of two radio lobes or sources on opposites sides of a central square with all three lying along a straight line is a feature that is seen in many quasars and radio galaxies. The structure suggests the occurrence of explosions, which throw energetic particles out from the central source along a preferred direction or symmetry axis. The possible nature of these explosions and their relevance to models for quasars is also discussed in the following section by Lightman.

In this same context, figure 8 shows the radio image (white contours) of the relatively nearby radio galaxy Centaurus A. The optical galaxy is shown as the large, fuzzy, dark spherical object. (The finite size and structure of this object in visible light make it a galaxy, whereas a quasar image in visible light is usually just a point source or dot.) Centaurus A also shows a double radio structure. We can learn more about the activity in

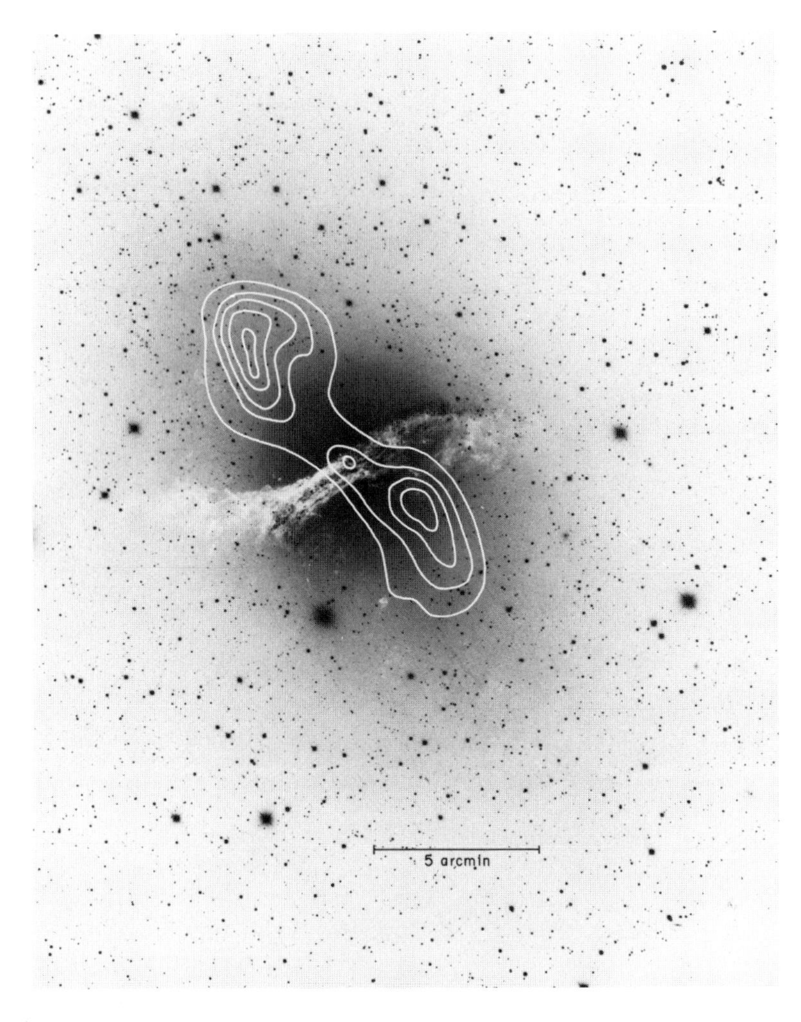

Figure 8
Radio contours (white) for Centaurus A superimposed on visible light photograph of galaxy (dark, fuzzy extended source in center). The radio map again shows a double-lobed structure. The scale of angular size is shown at the bottom, where 1 arcmin is one-sixtieth of one degree. (Radio contours from W. N. Christensen et al., *Monthly Notices Royal Astronomical Society* 181 [1976]: 183; optical photograph from Cerro Tololo Inter-American Observatory)

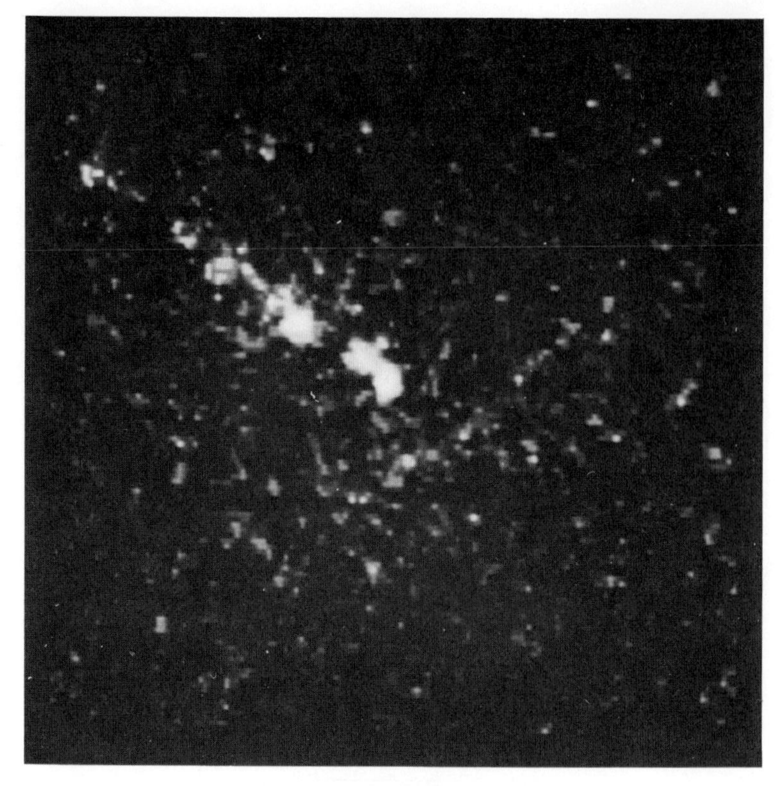

Figure 9
High-resolution X-ray image of Centaurus A showing central source associated with galaxy nucleus and jetlike structure extending from nucleus toward radio lobe. (Courtesy E. Feigelson and E. Schreier, Harvard-Smithsonian Center for Astrophysics)

this radio galaxy by studying a jetlike structure shown in figure 9. This figure is an x-ray image of Centaurus A obtained with the Einstein satellite. The central x-ray source is associated with the nucleus of the galaxy, but the striking feature is the jet which stretches out from the center in the direction toward one of the radio lobes. The picture indicates that energetic particles travel as a stream from the central source to the radio lobe. These particles may be the means of providing the radio lobe with the energy needed to continue radiating. The speed with which the matter in the jet is traveling allows us to estimate the age of the jet as 25,000 to 2.5 million years. The key feature is that the pointing of the jet from the nucleus to the radio lobe indicates that the central source remembers the direction of the explosion, and particles continue to follow a preferred path away from the central source. Jets similar to the one shown are seen in a number of quasars.

One more piece in the quasar puzzle comes from studying how the typical intensities of quasars change over times of millions and billions of years. Optical observations suggest that, in the past, either there were more quasars or the quasars were brighter than they are now. Because our x-ray observations show that most quasars are bright x-ray sources, we can combine these data with a measurement of the total output of all x-ray sources to set a limit on the number of faint quasars. Together the x-ray and optical data suggest that the main effect is a dimming of a quasar's brightness with time. Over some 10 billion years, roughly the estimated age of the universe, the average quasar brightness probably decreases by a factor of about one hundred.

The observational facets of quasars most relevant to theoretical interpretations can be summarized as follows:

1. Quasars are very luminous, often radiating one hundred to one thousand times more energy than an entire galaxy. Luminosities are determined from the observed intensities and distances calculated from red shifts.
2. Quasars are very small as deduced from their starlike appearance and the short times over which their intensities change.
3. Quasars do not radiate predominantly as hot gas, as stars do. The shapes of their electromagnetic spectra from the radio to the x-ray band suggest that the synchrotron and Compton scattering processes are involved.

4. Quasars are often sites of violent explosions, usually with preferred or remembered directions as indicated by the double-lobed radio sources and the jets observed in a number of sources.

The study of quasars requires combined data from radio to visible light to x-rays. Also, rather than be spurred by predictions, the field has tended to develop primarily through observational discoveries of quite unexpected phenomena. In turn, these observations are beginning to require the acceptance of ideas once considered quite radical.

Further Reading

Abell, G. *Exploration of the Universe.* New York: Holt, Rinehart, and Winston, 1964.

Burbidge, G., and Burbidge, M. *Quasi-Stellar Objects.* San Francisco: W. H. Freeman, 1967.

Giacconi, R. "The Einstein X-Ray Observatory." *Scientific American* (February 1980), p. 80.

——— and Tananbaum, H. "The Einstein Observatory: New Perspectives in Astronomy." *Science* (August 22, 1980), p. 865.

Mitton, S., ed. "Active Galaxies and Radio Galaxies," In *The Cambridge Encyclopedia of Astronomy.* New York: Crown, 1977.

Schmidt, M. "3C273: A Star-like Object with Large Red-Shift." *Nature* (March 16, 1973), p. 1040.

ALAN P. LIGHTMAN

TOO MUCH FROM TOO LITTLE?

Energy, the mysterious stuff that turns the natural universe and makes life possible, is nowhere more awesome than in the vast reaches of space.

One of the earliest sources of energy recognized by humankind was the Sun, which irradiates our planet with heat and light. Figure 1, which dates about 135 B.C., is a bas-relief of the Egyptian pharaoh Akhenaton holding out his hands to receive power from the Sun God Aten. To the Egyptians, the Sun was a mystical, mysterious being, an object of awe and wonder. For later civilizations, the Sun often remained just as mysterious; but for early scientists, at least, the mystery was what force powered and drove that magnetic ball of fire. In the late 1800s, it was believed that the energy source of the Sun might be chemical; in this view, the Sun could be compared to a giant coal-burning furnace. Unfortunately such a furnace would burn itself up in a mere one hundred thousand years, a period much shorter than the then-accepted age of the solar system. In 1920 Arthur Eddington correctly suggested that the Sun is powered by nuclear energy, and indeed the Sun and other stars are essentially huge nuclear fusion reactors in the sky.

When a star like the Sun expends its nuclear fuel, however, it collapses under its own weight and thereby produces an enormous explosion called a supernova. Here the source of energy has become largely gravitational, which is, in fact, the most pervasive source of energy throughout the universe.

Still another kind of energy powers the radio pulsars. First discovered in 1967, these objects are believed to be rapidly rotating compact bodies known as neutron stars that gradually convert their rotational energy into powerful electromagnetic radiation.

When we extend our gaze to truly cosmic distances, and survey those aggregates of billions of stars known as galaxies, we are astounded to see entire galaxies exhibiting violent activity that requires the presence of extraordinary and still unexplained powerhouses at their centers. But even more energetic

Figure 1
Bas-relief of the Egyptian pharaoh Akhenaton of the Eighteenth Dynasty, about 1350 B.C. (From the Cairo Museum)

than the most active galaxy—indeed the greatest cosmic powerhouses in the universe—are the quasars. Unlike the case of the Sun, supernovas, or pulsars, the power source of the quasars is still a mystery; however, we have important observational and theoretical clues to their energy source. And uncovering the secret of the quasars' energy is a good illustration of a scientific detective story still in progress.

Theorists often form their own views of which observational results are important and which are not. Let me summarize some of the key observational properties of quasars, presented in more detail by Harvey Tananbaum. The foremost observational fact about quasars, which may be deduced by combining their large distances from Earth with the amount of energy that we receive from them, is that quasars are extremely luminous. Table 1 compares the luminosity of a typical quasar with that of a normal galaxy in emitted radio waves, infrared radiation, visible light, and x-rays, using one galaxy power in visible light as the basic unit. In total, an average quasar is one hundred to one thousand times more luminous than an ordinary galaxy.

The large luminosities of quasars and other observed properties provide important clues to the nature of the power supply. Tananbaum has shown evidence that a quasar produces energy for at least several million years. By multiplying the observed luminosity (energy per unit time) of a quasar by a minimum estimate of its lifetime, we obtain a minimum estimate for the total amount of energy produced (table 2). The total amount of energy produced by a typical quasar over its active lifetime is the energy equivalent of 10 million times the mass of our Sun—that is, the amount of energy obtained from complete conversion into pure energy of a mass that large.

How much mass was needed to produce a quasar's energy? No mechanism is 100 percent efficient at converting mass into energy. For example, nuclear processes of the type involved in hydrogen bombs and in stars are only one percent efficient, so a quasar powered by nuclear energy would require a billion

Table 1
Luminosities of galaxies and quasars

	Radio	Infrared	Visible	X-Ray
Normal spiral galaxy	10^{-5}	10^{-1}	1	10^{-4}
Quasar	1	5×10^{2}	10^{2}	10^{2}

Table 2
Energy and mass requirements of a quasar

	Nuclear energy	Gravitational Energy
Efficiency	1%	10%
Minimum solar masses in quasar core	10^9	10^8

Note:
Total energy = Luminosity × lifetime
= (10^3 galactic luminosities) × 1 million years
= 10^7 solar rest mass energies.
Total energy = (Total mass required) × (conversion efficiency).

solar masses cycled through its central powerhouse. Gravitational processes near black holes are 10 percent efficient, so a quasar powered by gravitational energy could operate on ten times less mass. No known physical process is much more than about 10 percent efficient at converting mass to energy. Thus a minimum estimate of the amount of mass cycled through the center of a quasar is about 100 million times the mass of the Sun.

How large is the energy-producing region of a quasar? Unfortunately much of a quasar's radiation, especially the x-rays, infrared, and some of the visible light, comes from regions of the quasar much too small to show up as anything but a point of light. Thus we have no direct knowledge of the size of these regions. A simple and clever theoretical argument, however, can help answer the question. The key is the quasar's time variability, the fluctuations in intensity of detected radiation that Tananbaum discusses.

A simplified model of time variability is shown in figure 2. All objects, whether they be eardrums vibrating to the sound of a distant voice or quasars vibrating from colliding shock waves, continually exhibit fluctuations in response to changes in their environment. These fluctuations are what cause time variability in the luminosity of a quasar. Suppose now that an observer is monitoring the radiation from a region of size R and that at $t = 0$ the region suddenly undergoes a pulse or vibration (indicated by dashed lines in the figure) and then quickly settles back into its previous quiescent state. Light rays immediately take off toward the observer from all points of the region. Because all these light rays travel with the same speed, the light ray that left the near side of the object will always be ahead of the light ray that left the back side of the object. At some later

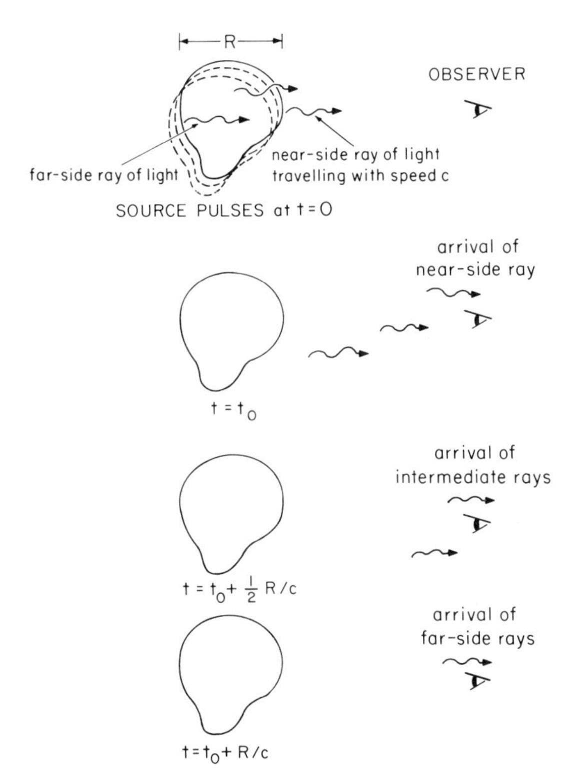

Figure 2
Simplified illustration of the detection of light rays from an object after it undergoes an instantaneous pulsation (indicated by dashed lines). After the initial fluctuation, the subsequent times of arrival of various rays are indicated under the object. Our detector at Earth is indicated by the eyeball. From a knowledge of the duration of the observed fluctuation, the size of the object R may be deduced. (Technically it has been assumed that the object is optically thin. The results are still roughly correct for optically thick objects unless they are relativistic.)

time, $t = t_0$, the near-side light ray arrives at the observer; but light rays continue to arrive so the light ray emitted from the middle of the region arrives next and, finally, the light ray emitted by the back side of the region arrives at the observer's telescope. After this last light ray arrives, the radiation observed from the region returns to its level before the fluctuation began. How long did this fluctuation last; that is, how long was the time between the reception of the first light ray and the last light ray? This interval of time must be just the time it takes the back-side light ray to catch up to the starting line of the near-side light ray. If we denote the speed of light by c, this time is R/c, or distance divided by speed. Therefore if the observer sees a fluctuation in detected radiation that lasts one year, then the size of the emitting region is determined to be one light-year in size, and so on. This indirect theoretical argument gives an estimate of the maximum possible size of an emitting region.

From the observed time variability in detected x-rays, for example, we can infer that the x-ray-emitting region is no larger than about a light-day across. A sober calculation then leads to the conclusion that a typical quasar can produce a hundred to a thousand times the luminosity of a normal galaxy from a region one hundred thousand times smaller in size. Roughly speaking, if the city of Boston were a galaxy in terms of its power output and size, then a quasar would have the power of the entire United States produced in a region the size of a baseball. Whatever powers the quasars must be highly efficient at converting mass to energy in order to produce so much energy from so small a volume.

Another important clue to understanding quasars is the strong symmetry exhibited by the ejected matter. Quasars seem to single out a particular direction in space and to remember this direction for most or all of their active lifetimes. Tananbaum has shown radio images of radio-wave-emitting blobs of matter extending in a line through the center of the quasar; there are also photographs in visible light and x-rays showing jets, or streams, of matter extending out from the center in straight lines. These strong symmetries may be clues to the nature of the power source at the center. Something within the quasar has managed to select out a special orientation, or symmetry axis, that has persisted for at least a million years. Furthermore this is not a freak occurrence, requiring fortuitous conditions. Many examples of quasars and highly energetic galaxies with such jets exist. Whatever causes this phenomenon must occur easily and commonly.

Finally the power source of quasars must be short-lived during the period of high activity. A typical quasar lifetime must be at least a million years. But there are some reasons why the highly energetic phase cannot be much longer. First, too long a lifetime would require the already enormous total energy required to be even larger. Second, we do not observe any quasars still highly active (because of the finite speed of light, this is equivalent to not observing any nearby quasars); all the quasars we observe went through their most energetic phase long ago. In other words, the energy supply mechanism of the quasar must exhaust itself rapidly, maintaining maximum intensity for only about 1 million to 10 million years.

The possible theoretical models that satisfy the constraints are listed in table 3, where a number of the more conservative models for the quasar power source proposed since 1965 are given. I say "conservative" because it has been suggested that the prodigious power of quasars can be explained only by new laws of physics. However, most researchers recoil at this idea. The explanation of quasar phenomena in terms of known energy sources is far preferable to the inventing of new laws of nature or highly exotic processes in an ad hoc manner. (Most scientists prefer simplicity whenever possible. One of the most beautiful and curious features of the natural world is that it seems to conform to humanity's own idea of mathematical and physical simplicity.)

Independent evidence shows that the quasar phenomena do not require a qualitatively new form of energy or energy release mechanism. If we order the different types of galaxies according to their luminosities, we find a continuous sequence going from normal galaxies like our own, to radio galaxies like Centaurus A, to highly energetic active galaxies like M87. (See figure 3.) As it turns out, the most energetic of the active galaxies, those known as Seyfert galaxies, are about as bright as the least luminous quasars. All of the energy sources listed in table 3 are related in that they utilize gravitational energy in some form. Indeed it is possible that some of these processes form an evolutionary sequence.

The last model, gas flow onto a massive black hole, was first proposed as the quasar energy source in 1969 by Donald Lynden-Bell of England. This model is the most efficient mecha-

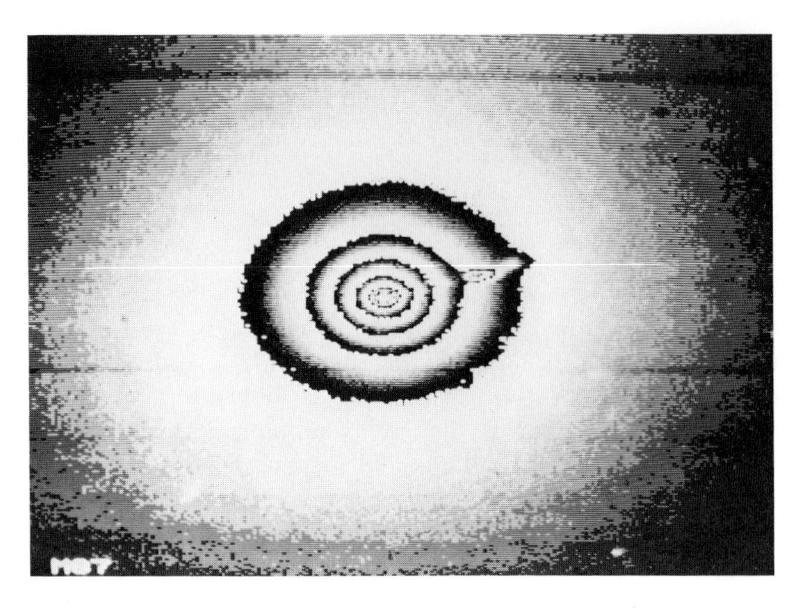

Figure 3
Electronic detector photograph of active galaxy M87 showing jet. (Photograph from Smithsonian Astrophysical Observatory)

nism for converting mass to energy and could be the correct explanation for the most energetic quasars.

A black hole is a region of space created by the total gravitational collapse of matter whose attractive gravitational force is so intense that no matter, light, or communication of any kind can escape. Black holes are believed to be formed in the dying phase of stars larger than about three times the mass of the Sun. During most of its lifetime a star supports itself against its own inward gravitational attraction with outward thermal pressure. This pressure is generated by nuclear reactions in the deep stellar interior. But every star must eventually deplete its nuclear fuel. With its gravitational pull unbalanced, the star will begin collapsing. For a sufficiently massive star, the collapse

Table 3

Models for quasar power source
1. Random succession of gravitational collapses and supernovae
2. Collision of stars in dense star clusters
3. Gravitational collapse of single massive object
4. Rapidly rotating massive object (spinar)
5. Gas flow (accretion) onto massive black hole

cannot be halted. As the star collapses to a smaller and smaller size, its gravity becomes stronger and stronger until the escape speed from its surface reaches 186,000 miles per second, the speed of light. (The escape speed is the speed required by a projectile to escape the gravity of a planet, star, or other body. The escape speed from Earth is about seven miles per second.) Once this point is reached, the star essentially cuts itself off from the surrounding universe, and a black hole is formed. The size of a black hole is directly proportional to its mass; a black hole of the mass of the Sun would be a sphere about 3.5 miles in diameter.

Significantly, the concept of a black hole was not invented to explain the quasar phenomenon. Black holes are a consequence of Einstein's 1915 theory of gravity, General Relativity, one of the fundamental bases of modern theoretical physics. In 1939 J. R . Oppenheimer and H. Snyder were the first to discover black holes theoretically by manipulation of Einstein's equations. Yet for the next twenty-five years black holes were not taken seriously because of their bizarre properties. In fact, black holes were still considered highly speculative at the time quasars were discovered in 1962.

In 1964 the American astrophysicist E. E. Salpeter and the Soviet astrophysicist Ya. B. Zeldovich independently suggested that black holes could be highly efficient cosmic powerhouses under appropriate conditions. Their idea was that a black hole in space may be surrounded by gas or other matter. If so, this matter will be drawn toward the hole by gravity. As the matter moves toward the black hole, it falls faster and faster, converting its gravitational energy into energy of motion, heat, and ultimately radiation, which we can detect. Just before the matter falls into the black hole, as much as one-tenth of its mass will have converted into radiation energy. This idea of utilizing gravitational energy by matter falling toward a gravitating body was a simple one and well known in more mundane circumstances. If we drop an eraser, the eraser falls, going faster and faster until it strikes the floor, converting its gravitational energy into motion and then into a little heat upon impact. If we dropped the eraser from very high up, the total amount of gravitational energy converted into heat would be about one-billionth of the eraser's mass. Exactly the same process occurs when matter falls toward a black hole; the only difference is that the gravity of the black hole is much stronger than the

Earth's so that a much larger fraction of the mass of the falling matter can be converted into energy.

The insight provided by Salpeter and Zeldovich was that this simple gravitational process can occur on cosmic proportions with black holes. It is always fascinating when a certain theoretical idea or result is published by two different scientists within a period of months, completely independently. And frequently this happens when there were no new observational results that could have been a catalyst. It is as if there were a kind of ripeness in the air for a particular theoretical development.

Soon after this prediction of black holes as cosmic powerhouses, pulsars were discovered in 1967 and then the first good candidate for a black hole, Cygnus X-1, in 1971. Black holes became more acceptable as an astrophysical possibility.

Having attained some scientific respectability, black holes have been incorporated into a theoretical model for quasars in the following way. A large total amount of mass is required to produce the energy output of quasars—from 10 million solar masses for a weak quasar to 1 billion solar masses for a strong quasar. Because this large mass must be confined to a small space, gravitational collapse, with formation of a black hole, seems inevitable. To release its gravitational energy, this mass must be captured by the black hole. Thus the black hole itself must either initially or eventually attain a mass this large. In terms of the underlying theory, such massive black holes are no more difficult to write equations for and to conceptualize than stellar-sized black holes like Cygnus X-1. Massive black holes could be formed in the gravitational collapse of gas during the initial formation of a galaxy, or as a result of the collision and coalescence of many stars, or as a result of a small black hole's capturing matter and growing.

One of the attractive features of the black hole model for quasars is the high efficiency of conversion of matter to energy. Matter falling onto a black hole can convert more than 10 percent of its rest mass into radiation.

It is a simple matter for a black hole to produce a symmetry axis, as required by the observations. It simply has to rotate, a state that will develop naturally if the gas forming the black hole and falling into it has a rotation of its own. This condition would naturally be expected, since most large aggregates of matter like our own solar system and even whole galaxies are observed to have a net rotation. A rotating black hole affects matter along its axis of rotation in a very different manner from that in its equatorial plane, just as weather patterns near the north pole of the Earth are very different from those near the equator. The rotation axis of the hole then becomes the preferred direction in space. Since the rotating hole is like a big gyroscope, this axis is quite stable and can hold its direction for a long time. A number of possibilities for collimating gas into the striking jet structures are now being pursued. These ideas differ in detail, but all hinge on the fundamental property of a collective rotation. In contrast, any mechanism that invokes a succession of different energy releases, scattered randomly throughout a large region, would probably involve a range of spatial orientations and would have difficulty in producing a single, well-directed jet of matter.

Figure 4 illustrates the type of configuration that might be expected. At the center is a black hole, deformed into a partially flattened shape by its rotation. A cross-sectional view shows gas flowing toward the black hole and forming a bloated doughnut shape. A variety of mechanisms exist by which matter might be funneled out in both directions along the rotation axis of the hole. This is shown in the figure as the outwardly directed lines. The beamed matter then can be further collimated by cylindrical magnetic fields, shown as loops in the figure. Some of the energy may be supplied by the rotation of the black hole, analogously to the radio pulsars. This highly directed form of energy is particularly suited for producing matter streams traveling at nearly the speed of light (relativistic matter streams). Much of the observed x-ray radiation may be produced by the gas in the doughnut configuration around the hole, heated to temperatures of 100 million degrees or higher.

To produce energy the black hole must have a supply of gas and matter. Where does the material originate? Because we know the rate at which energy must be produced and the efficiency of a black hole for converting mass into energy, the rate at which matter must be fed to the black hole can be calculated. For a quasar of moderate luminosity producing a hundred galaxy luminosities, the production of gas in an ordinary galaxy by the aging and explosion of stars is sufficient to fuel the black hole. This mechanism generates gas at the rate of about 1 solar mass per year. For the more luminous quasars, which produce a thousand or ten thousand galaxy luminosities, additional

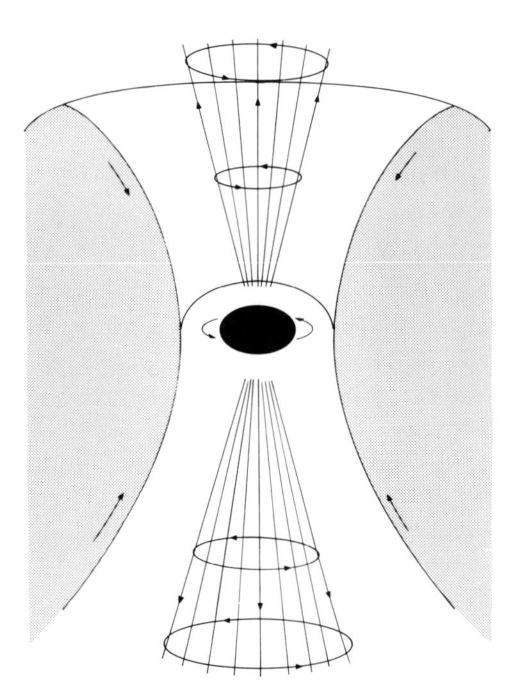

Figure 4
Hypothetical picture of a rotating black hole accreting gas and expelling a jet of matter along its rotation axis. This is a possible model for the power source of quasars. Gas flowing inward forms a bloated doughnut shape (cross-sectional view is shaded in the figure) and releases its gravitational energy. The outgoing jet of matter (indicated by diverging lines) may be partially confined by cylindrical magnetic fields.

mechanisms of gas supply must operate. Two such mechanisms are the physical collisions of stars with each other and the ripping apart (tidal disruption) of stars that pass too close to the black hole. In each case, gas is liberated from the destroyed stars and falls toward the black hole. Both of these mechanisms are effective when the black hole is surrounded by a dense system of stars. On a small scale, such dense star systems have been observed in our own galaxy as the beautiful globular clusters, aggregates of about 100,000 stars that orbit each other and clump together under their mutual gravitational attraction. (See Huchra's figure 1.)

The tidal disruption mechanism is illustrated in figure 5, showing a system of stars surrounding a black hole. For purposes of the illustration, we have assumed that the black hole has a mass of about 100 million times the mass of the Sun and

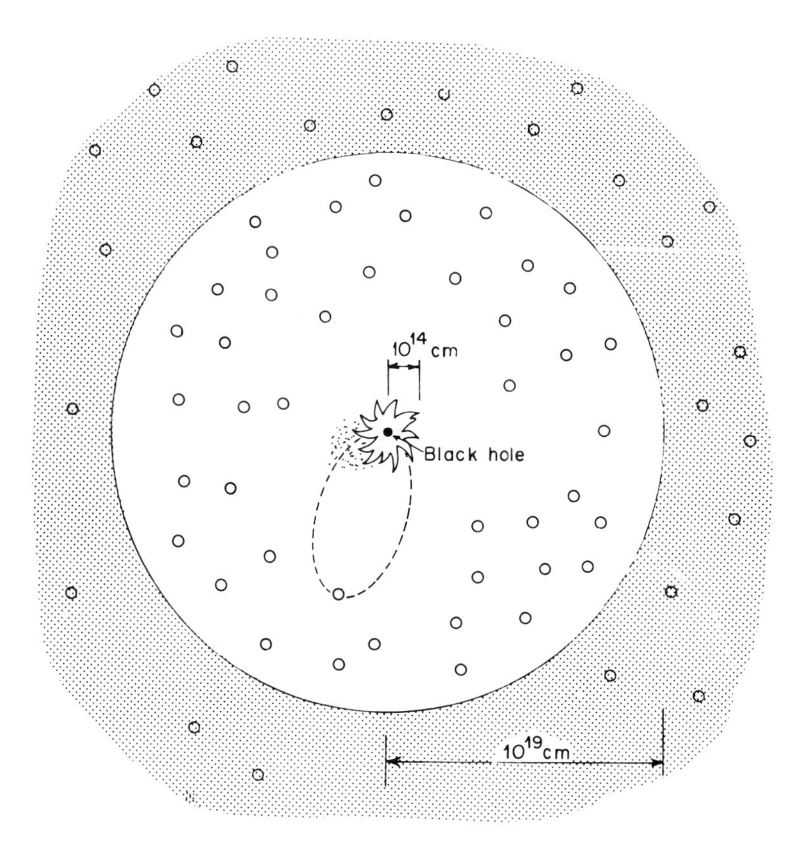

Figure 5
The tidal disruption of stars by a massive black hole. All stars in the shaded region are unaffected by the black hole. The remaining stars orbit the hole, gradually move toward it, and are gravitationally ripped apart when their orbits approach too close to the hole (shown by the buzz saw in the figure). The gas from destroyed stars may be the primary fuel supply to power the black hole.

that stars far from the black hole are traveling at a speed of about 300 miles per second, not a large speed for the centers of galaxies. Stars more than 10^{19} centimeters (about ten light-years) distant from the black hole will be unaffected by the presence of the hole. This is the shaded region of the figure. Stars closer to the hole will be strongly gravitationally attracted to it, will orbit the hole, and will accumulate in this region. Any star passing within about 10^{14} centimeters (about ten times the distance from the Earth to the Sun) of the hole will be ripped apart by the tidelike stretching of the hole's gravitational field, somewhat analogous to the ocean tides caused by the Moon. The tidal region of destruction is indicated in the figure by a buzz saw around the black hole.

Both the stellar collision and the tidal disruption mechanisms for gas supply have the desirable feature that they eventually exhaust themselves by destroying all of the available stars in the vicinity of the black hole. For example, Centaurus A, the radio galaxy, may have once been a quasar, but now houses at its center a quiet black hole that has exhausted most of the available fuel. Different mechanisms of gas supply each cause a different long-term evolution of the quasar luminosity with time. In principle, comparison of theory and observation for such evolution could shed some light on the nature of the gas-supply mechanism.

I have addressed only some of the observational data on quasars discussed by Tananbaum. Despite the wealth of detailed observational results available, our understanding of the energy source is based on only the simplest and broadest theoretical ideas, like energy and mass requirements, efficiencies, and so on. Does this speak for the complexity of the phenomenon, or perhaps a limited viewpoint that has overlooked some important clues? Are we trying to get too much from too little?

In another sense, it is a triumph that we, little cosmic ants on our little planet Earth, understand as much as we do about quasars, objects clear across on the other side of the visible universe, and black holes, for fifty-five years objects existing only as equations on a piece of paper.

Further Reading

Brecher, K. "Active Galaxies." In *Frontiers of Astrophysics,* edited by E. H. Avrett. Cambridge: Harvard University Press, 1976.

Dyson, F. J. "Energy in the Universe." *Scientific American* (September 1971).

Lightman, A. P. "Black Holes." In *Collier's Encyclopedia.* New York: Macmillan.

Morrison, P. "Resolving the Mystery of the Quasars." *Physics Today,* (March 1973).

Rees, M. J. "Quasar Theories." *Annals of the New York Academy of Science.* 302 (1977): 613.

Ruffino, R., and Wheeler, J. A. "Introducing the Black Hole." *Physics Today* (January 1971).

Schmidt, M., and Bello, F. "The Evolution of Quasars." *Scientific American* (May 1971).

CHAPTER 6

WALTER H. G. LEWIN

THE MYSTERY OF THE X-RAY BURST SOURCES

Scientists sometimes call their unsolved problems mysteries. This word is reminiscent of detective stories full of excitement and surprises that are brought to a dramatic solution through the careful and clever piecing together of a puzzle. There are indeed many such stories in science. One of them is the mystery of the cosmic x-ray burst sources.

These so-called x-ray bursters typically emit ten times more visible light than our Sun does. That does not make them particularly special, as one might gather from the previous description of quasars. However, they emit over a billion times more x-radiation than does the Sun. In addition to this strong and somewhat variable continuous flow of x-radiation, every so often (a few times a day) these objects produce a short burst of x-rays. The bursts reach their maximum intensity in only a few seconds, sometimes even faster. The bursts typically last a minute, though much shorter and much longer bursts have been observed. At burst maximum, the x-radiation is about ten to one hundred times stronger than the continuous x-radiation between bursts. The x-ray burst sources are very powerful indeed; it takes the Sun about two weeks to generate the same amount of energy that one of these sources can produce in a single brief x-ray burst.

What kind of an object is capable of this exotic behavior, and how does it produce the x-ray bursts? These two questions have occupied and challenged us for several years. It was a puzzle indeed. At first a few pieces were put in the wrong place, blurring our vision, but as the pieces fell in place, one by one, a clear picture emerged.

The story begins in the early 1970s when about one hundred variable galactic x-ray sources were known. Most of these were believed to be binary systems (a system consisting of two stars in orbit about each other) in which one member is an x-ray source such as a neutron star or, in a few cases, a black hole (see chapter 5, Alan Lightman) and the other member is a normal companion star. (Normal here indicates that the companion star is burning its nuclear fuel, as our Sun does; the other object, be it a neutron star or a black hole, has already exhausted its nuclear fuel supply.) The two stars are very close to each other and are therefore called a close-binary system.

The neutron star in such a system has a mass comparable to that of the Sun (1 solar mass) but a radius of only about 10 kilometers (the Sun's radius is about 700,000 kilometers) and thus has an extraordinarily high density of about 10,000 billion grams per cubic centimeter. Neutron stars are of such high densities that their atoms are crushed into almost pure neutrons; hence the name of the star. The gravitational field strength near its surface is about 100 billion times higher than the gravitational field near the Earth's surface. Matter, mostly hydrogen, can spiral from the surface of the nearby normal star toward the neutron star and, in doing so, will form a disk surrounding the neutron star; this is called an accretion disk. When this matter approaches the neutron star, it will reach very high velocities due to the strong gravitational field of the neutron star, and consequently it will heat up to temperatures of about 10 million to 100 million degrees. At these very high temperatures this matter will radiate predominantly x-rays.

In 1971 using the first orbiting x-ray observatory UHURU, Riccardo Giacconi, Ethan J. Schreier, Harvey D. Tananbaum, and their coworkers discovered periodic pulsations (with periods of a few seconds) and eclipses (with periods of a few days) in the x-radiation from two strong x-ray sources. The following picture emerged. The x-ray pulsations result from the rotation of a strongly magnetized neutron star whose magnetic dipole axis is not aligned with the star's axis of rotation. Due to the strong magnetic field the matter cannot fall freely onto the neutron star but is forced to fall only near the magnetic poles. This results in two hot, highly radiating regions on the neutron star (the magnetic poles), which rotate about the star's axis of rotation, thereby changing their direction in space. In other words, everytime a hot spot points toward the Earth we see a pulse of x-rays. The x-ray eclipses occur when the neutron star disappears behind the much larger normal companion star.

After this important discovery, it was believed that the majority of the variable x-ray sources in our galaxy would be similar to the above. To verify this assumption, many strong and variable x-ray sources were carefully studied to search for their binary nature and to find x-ray pulsations, which would establish that the x-ray star is a neutron star rather than a black hole. (Black holes have no surface nor is there a strong magnetic field associated with them; therefore they cannot produce periodic x-ray pulsations.) Surprisingly, the search turned up a large number of x-ray sources that showed no signs of a binary nature and from which no x-ray pulsations were observed. What would these sources be? Could they be fundamentally different from the classic x-ray binary systems? As early as 1975, there was much exciting speculation. Before discussing this speculation, I will briefly explain the various methods by which the binary character of x-ray systems can, in principle, be established. This will be of use later to appreciate how our ideas evolved.

The observation of periodic x-ray eclipses leaves no doubt about the binary nature of an x-ray source. If eclipses are not observed, however, it does not follow automatically that the x-ray star is not in a binary system. The absence of evidence is not the evidence of absence. X-ray eclipses will occur only if we, on Earth, are viewing the orbital plane of the binary system almost edge-on. If, by contrast, we are viewing the orbital plane at a sufficiently large angle, eclipses will not be observed. In the absence of eclipses, the binary character of a source can in principle be found by observing the Doppler effect (see the discussion by Harvey Tananbaum) in the x-ray pulsations, if such pulsations are observed. If x-ray pulsations are absent, optical observations (at visible light rather than x-ray wavelengths) may reveal the binary nature. If the normal companion star in an x-ray binary system is bright enough, it can be studied from the ground with optical telescopes. If it turns out that the visible star is a spectroscopic binary (the colors shift periodically due to the orbital Doppler effect), the binary nature is beyond question. Thus several lines of investigation can be followed to uncover the binary nature.

Using the technique I have described, several more binary systems were discovered, but nowhere near as many as expected. In most binary systems, the normal companion stars were very bright, and their masses turned out to be ten to twenty times larger than that of the Sun. I will call these binary systems class 1 objects. The mass of the companion star, not the presence of x-ray eclipses or x-ray pulsations, determines whether an x-ray source belongs in this class. X-ray eclipses are often observed in class 1 objects, but this depends on the orientation of the binary system; x-ray pulsations are almost always observed in class 1 objects, but there are a few exceptions. There is an x-ray binary system, called Cygnus X-1, in which the x-ray star is probably a black hole. Its companion star is very massive (it is a class 1 object), but no x-ray eclipses and no x-ray pulsations are observed from this source.

The class 1 objects must be relatively young, since stars with a mass about twenty times that of the Sun do not live very long. They burn up their nuclear fuel in only a few million years, after which they die abruptly in a supernova explosion. This explosion ejects matter into interstellar space, and it can leave as an additional remnant a small and dense object (a neutron star). Since the class 1 objects contain a very massive nuclear-burning star, they must be younger than a few million years.

The large number of variable x-ray sources that showed no signs of binary nature and from which no x-ray pulsations were observed were never associated with bright stellar objects. Interestingly, nearly all sources in this class (I will call them class 2 objects) are located in the general region of the galactic center and in globular clusters. This implies that they are among the oldest objects in our galaxy; they are probably about 10 billion years old.

This was more or less the situation in 1975. The class 1 objects were reasonably well understood, but the class 2 objects were a complete enigma. What were these class 2 objects? Could they be single black holes? The absence of x-ray pulsations and eclipses could then be explained because black holes cannot produce x-ray pulsations. Furthermore if a black hole is not orbiting a companion star, one would not observe x-ray eclipses, and the absence of a bright, visible companion star would then also be understood. There is, of course, the possibility that the class 2 objects are single, nonpulsating neutron stars. Nonpulsating neutron stars may well exist because x-ray pulsations would not be expected from neutron stars with weak magnetic fields or from those whose magnetic dipole axis is coaligned with the star's rotation axis.

A single black hole or a single neutron star would have to be surrounded by a cloud of gas so that it can accrete (gravitationally capture) a sufficient amount of matter to account for the observed x-ray luminosity (see the discussion by Alan Lightman). Typically an accretion rate of about 10^{17} grams per second is required. (At this rate, it would take two thousand years for a mass comparable to that of the Earth to accrete.) A black hole or a neutron star of about 1 solar mass placed in the very low density of interstellar space would not accrete at the required rate. However, the higher its mass and the higher the density of its surrounding gas, the higher will be its accretion rate. Thus a very massive but small object in a low-density (but not so low as interstellar space) environment could perhaps do the job. It is believed that no neutron stars in excess of about 3 solar masses exist, but the mass of a black hole is unlimited. Thus a very massive single black hole placed in a low-density environment was considered as a possibility. In fact, in 1975 John N. Bahcall and Jeremiah P. Ostriker and, independently, Joseph I. Silk and Jonathan Arons proposed that black holes in excess of 100 solar masses could be responsible for the x-radiation from those globular clusters that contain an x-ray source, a truly fascinating possibility.

The single-black-hole idea gained some additional support late that same year when Jonathan E. Grindlay and John Heise, observing with the Astronomical Netherlands Satellite (ANS), discovered so-called x-ray bursts from a class 2 object located in the globular cluster NGC 6624 (figure 1). X-ray bursts were also independently discovered by Richard Belian, Jerry Conner, and W. Doyle Evans. Grindlay and Herbert Gursky argued that the bursts observed with the ANS might be caused by a black hole in excess of 100 solar masses. They did not propose a specific mechanism for the bursts but believed that the observed variability in time could be accounted for by a scattering of primary x-rays in very hot gas of about 1 billion degrees. Such a very hot gas would escape the gravitational pull of solar-mass-sized objects but not a black hole with a mass in excess of a few hundred solar masses. Grindlay and Gursky also suggested that the bursts originated near such a very massive black hole. In the light of the earlier and fascinating black-hole model for x-ray sources in globular clusters, it was perhaps tantalizing to push this exotic idea. Unfortunately it turned out to be incorrect.

Following the discovery of x-ray bursts, our group at MIT (using the SAS-3 x-ray observatory) discovered in February and March 1976 five more x-ray burst sources, and, before the year was over, various groups observing with different satellites (Vela, OSO-8, UHURU, and SAS-3) found an additional ten burst sources. All of these burst sources were class 2 objects, and in spite of all efforts not a single burst was observed from

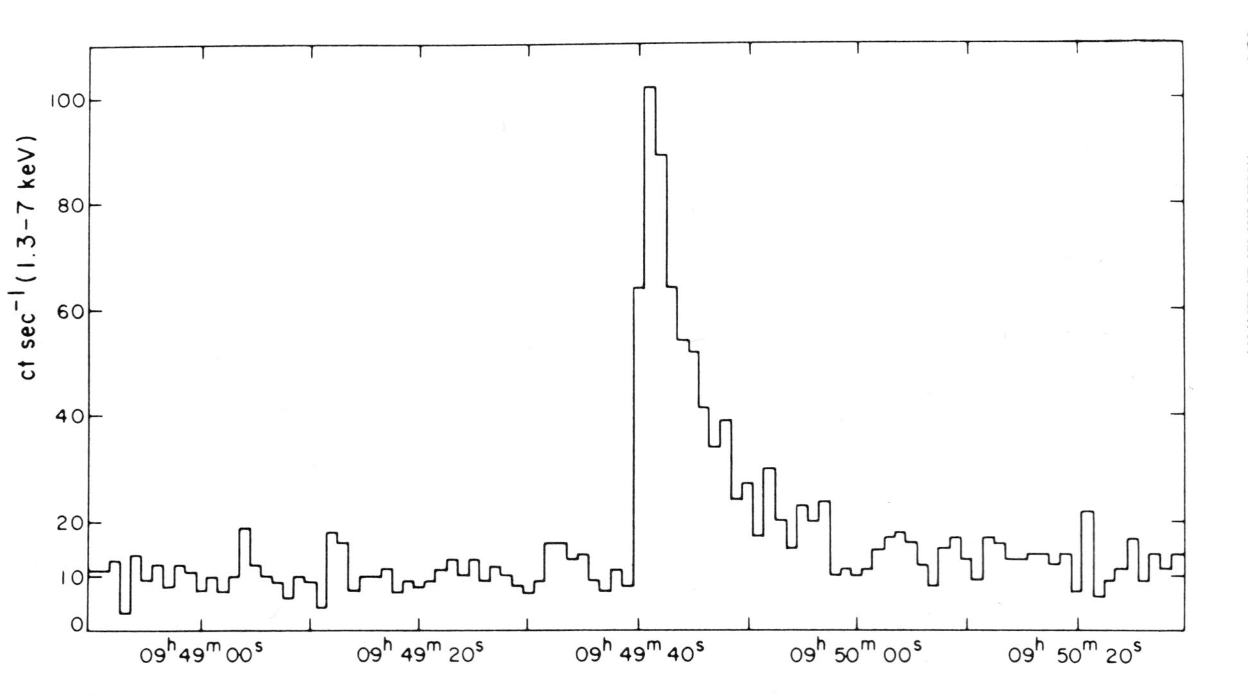

Figure 1
Discovery of an x-ray burst from a source located in the globular cluster NGC 6624 on September 28, 1975, with the Astronomical Netherlands Satellite. The vertical axis indicates the x-ray intensity in terms of the number of x-ray events (counts) recorded per second by the x-ray detectors in the 1.3- to 7-keV energy (or frequency) band. (Visible light has a frequency corresponding to about 0.001 keV.) The time is indicated along the horizontal axis (h denotes hours; m denotes minutes; s denotes seconds). Notice the fast rise of the burst in about 2 seconds. (Figure from Grindlay et al., *Astrophysical Journal* 205 [1976]: L127.)

a class 1 object. Clearly the bursts were unique to the class 2 objects (even though not all class 2 objects produce x-ray bursts), and it occurred to some of us that x-ray bursts might hold the key to the secrets of the entire group of class 2 objects. Moreover it became clear in 1976 that x-ray stars that produce bursts are ordinary neutron stars. A description of the early developments that led to this first important piece of the puzzle follows.

Jean H. Swank and her coworkers, observing with the OSO-8 satellite, were the first to notice that the x-ray spectrum (amount of energy in each color band) of one particular very long burst of about 600-second duration resembled that of a black body. (Every object absorbs a fraction of the radiation incident on it and it emits radiation. An object is called a black body when all radiation incident on it is absorbed. Such an object emits radiation that has a specific spectrum, determined only by its temperature. The total power emitted by a black body is proportional to its temperature to the fourth power and to its surface area.) During the first sixty seconds, the temperature rose to about 26 million K, then decreased, and, after about one hundred seconds, the temperature reached about 15 million K. The object continued cooling over time. The energy flux measured at the Earth is inversely proportional to the square of the distance between the Earth and the x-ray source. If the distance to the source is known, and the source's surface temperature has been measured from its x-ray spectrum, its total power and its surface area (and radius) can be calculated. For an assumed distance to the star of 30,000 light-years, Swank and her colleagues found a radius of about 100 kilometers during the first fifteen seconds of the burst and thereafter a more or less constant radius of about 15 kilometers. Incidentally the assumed distance was not unreasonable since most burst sources lie near the galactic center (figure 2) at a distance from Earth of approximately 30,000 light-years. Following this example, Jeffrey A. Hoffman, John P. Doty, and I established that the radii of two other x-ray stars that produce bursts were both about 10 kilometers at all times during the cooling period. We were unable to measure the radius of the x-ray star during the burst rise, when the temperature was still going up.

These radius measurements provided the first persuasive evidence that the bursting x-ray stars, and therefore probably the

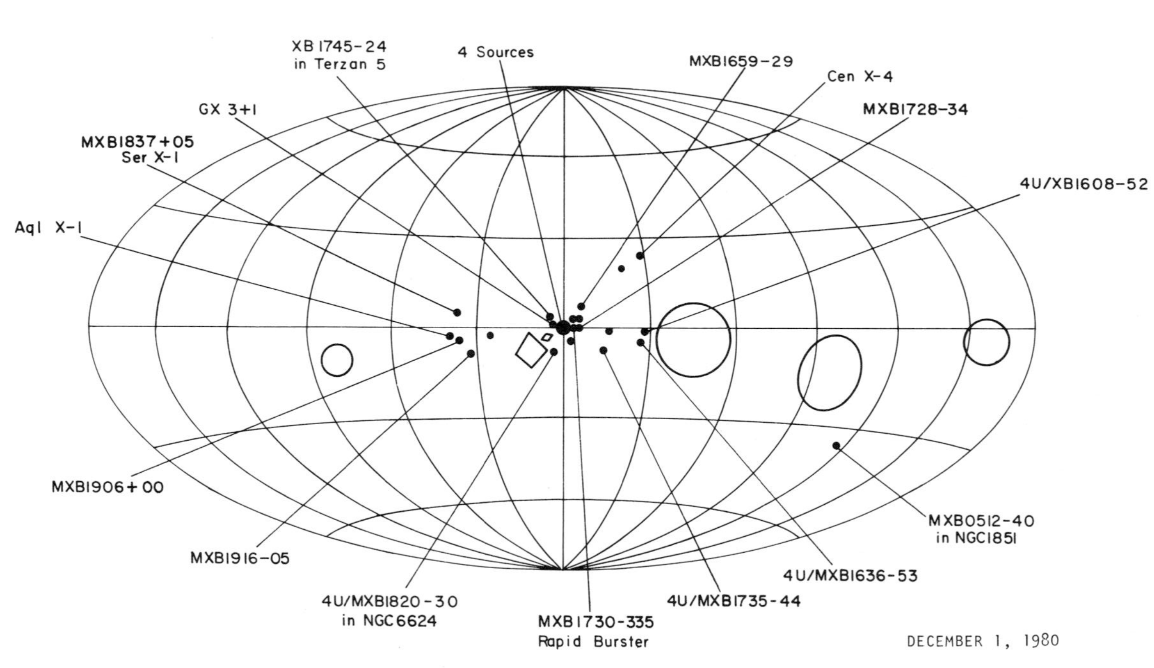

Figure 2
Sky map (in galactic coordinates) of thirty-one x-ray burst sources based on data available December 1, 1980. The circles and diamonds indicate regions in which a burst source is located. Several source designations are shown. The burst sources are concentrated near the galactic center (at a distance of about 30,000 light-years from us). About eight of them and possibly more are located in globular clusters. This is strong evidence that the burst sources, which are a subset of the class 2 objects, are very old. (Figure from W. H. G. Lewin and P. C. Joss, 1981 *Space Science Reviews*)

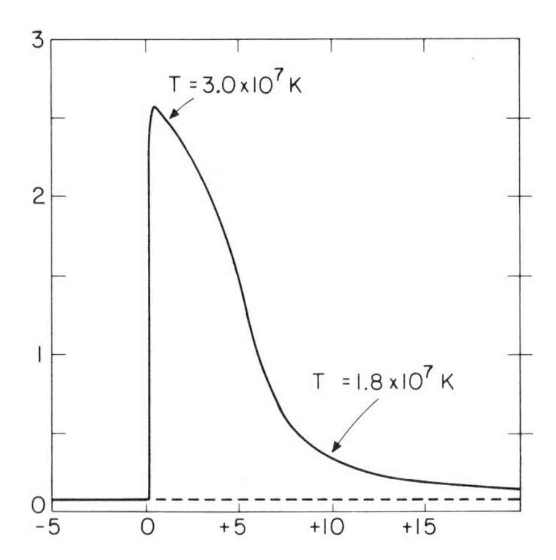

Figure 3
Theoretical x-ray burst. This is the result of one of the thermonuclear flash model calculations performed by Paul C. Joss. The vertical axis is the x-ray intensity in relative x-ray units; the horizontal axis is in seconds. Notice the fast rise of the burst. The intensity at the peak of the burst is about twenty-five times the preburst x-ray intensity. Compare the burst profile with those of figure 1; the agreement is quite good. Near the maximum of the burst the temperature of the neutron star's surface is about 30 million degrees. About ten seconds later the temperature is about 18 million degrees. (Figure from P. C. Joss, *Astrophysical Journal* 225 [1978]: L123)

majority of the x-ray stars in class 2 objects, are neutron stars. If the burst sources were very massive black holes, one would expect to find considerably larger radii because the radius of a 200-solar-mass black hole is approximately 600 kilometers. (See chapter 5, Alan Lightman.) Jan van Paradijs showed in 1977 that the small radius of about 10 kilometers is a property shared by all x-ray stars that produce bursts.

Further evidence for the presence of a neutron star in class 2 objects came in the spring of 1978, when the first thorough theoretical model calculations, made by Paul C. Joss, showed convincingly that the observed x-ray bursts can be produced by a nuclear blast on the surface of a neutron star (figure 3). The principle of the model is simple. Hydrogen accretes onto a neutron star. This accretion leads to the persistent emission of x-rays, which is derived from gravitational potential energy just as in the class 1 objects. The hydrogen accumulates on the neutron star's surface and fuses steadily to helium (the helium layer formed this way is below the hydrogen layer). After about

10^{21} grams have been accreted, the temperature and density in the helium layer can become critical, and the fusion of helium to carbon can occur rapidly in a so-called thermonuclear flash. This very powerful flash produces an x-ray burst. The intervals between the bursts depend on the mass flow onto the neutron star and on the temperatures in the deep interior of the neutron star. For a typical accretion rate of about 10^{17} grams per second, the intervals between bursts would be about three hours; for lower rates, the intervals would be correspondingly longer. The thermonuclear flash model has been quite successful in explaining many of the observed features. For instance, it explains the fast burst rise of about one to a few seconds, the observed intervals between the bursts, the observed temperature of about 30 million degrees at burst maximum, and the subsequent cooling of an object with a radius of about 10 kilometers (a neutron star). Moreover, the thermonuclear flash model, if correct, excludes black holes, since black holes have no surface on which the accreted material can accumulate and flash.

During periods of high burst activity, the energy radiated in x-ray bursts from a given source is observed to be about 1 percent of the energy radiated in the persistent x-ray emission. This can be easily understood in terms of the thermonuclear flash model. Each proton (a proton is a hydrogen nucleus) that falls onto the surface of a neutron star gains kinetic energy equal to about 10 percent of its mass, and this energy is released in the form of the persistent x-ray emission. (See the discussion by Lightman.) The hydrogen fuses steadily to helium, and the helium fuses to carbon in a thermonuclear flash. In the fusion of helium to carbon, the nuclear energy released per proton is only about 0.1 percent of the proton's mass. It therefore follows that, averaged over some time, the energy in bursts should be about a hundredth of the energy released in the persistent flow of x-rays. In short, the released nuclear energy in the burst is about 1 percent of the gravitational potential energy released in the accretion of matter onto the neutron star.

If many class 1 and class 2 objects contain a neutron star, why then are there such striking differences between these two classes? Many specific questions come to mind. Why do only class 2 objects burst (not all of them burst) and why do they not pulse; and why do only class 1 objects pulse (not all of them pulse), and why do they not burst? Other important questions also need answers. Are class 2 objects binary systems like the class 1 objects, and, if so, why do we not see a bright optical counterpart as we do in the case of the class 1 objects? Why were no x-ray eclipses seen in twenty class 2 objects carefully studied? Let us answer these important questions one at a time.

The reason why the class 2 objects burst and do not pulse and why the class 1 objects pulse and do not burst is related to their enormous age difference. Young neutron stars, often found in class 1 objects, can have very strong magnetic fields, which are responsible for the x-ray pulsations. The magnetic field in the very old neutron stars that we find in class 2 objects presumably has decayed away. The accretion is therefore not confined to the magnetic poles, and thus no beaming mechanism exists to produce x-ray pulsations. This explains why class 1 objects but not class 2 objects, pulse.

It is more difficult to see why class 2 objects, but not class 1 objects, burst. Joss has suggested that this too is related to the suspected difference in magnetic field strength in the two classes of objects. The concentration of the accreting matter onto the magnetic polar caps and the intense magnetic fields in the surface layers of the neutron star both tend to make the nuclear fuel burn more steadily rather than in distinct flashes. Thus the difference in magnetic field strength of the neutron stars in class 1 (young objects) and class 2 (old objects) is probably responsible for the differences in the two types of objects.

Another key issue is the binary nature of the class 2 object. There is no doubt that the class 2 objects are binary systems. We know that the neutron star accretes about 10^{17} grams per second. Only a nearby gas container (a star) could supply this. (Other proposals have been made, but they are inconsistent with the observations.) We are left with two important questions: why do we not observe x-ray eclipses from class 2 objects, and why do we not see bright optical companion stars as observed in class 1 objects? These questions go to the heart of the problem: the difference between the companion stars in the class 1 and class 2 objects.

Edward P. J. van den Heuvel suggested in 1977 that the companion stars of class 2 objects may have a mass of only about 1 solar mass, significantly less than that of the class 1 objects. Subsequently Joss and Saul A. Rappaport suggested that the mass of the companion stars of the class 2 objects may be even less than 0.5 solar mass. Such low-mass objects may be the answer to both the above questions.

Such a low-mass companion star would not be very luminous, and it would be difficult or perhaps even impossible to observe such a star when it is about 30,000 light-years from us. For example, if we placed the Sun at a distance of 30,000 light-years, it would appear as a very faint star that could be detected at best only with the largest optical telescopes.

It is more difficult to explain simply why a low-mass companion star would prevent the observation of eclipses, but this is an important piece of the puzzle. The probability that the neutron star, in a randomly oriented binary system, will be eclipsed by its companion depends only on the size of the companion star and on the distance between the two stars. The radius of the nuclear-burning companion star depends on its mass. In going from very low-mass stars of about 0.01 solar mass to stars of about 0.08 solar mass, the stellar radii decrease from about 60,000 kilometers to about 30,000 kilometers. For stars in excess of about 0.08 solar mass, the radii increase with increasing mass; a 1-solar-mass star (our Sun) has a radius of about 700,000 kilometers. It is sufficient for our purpose here that the radius of a nuclear-burning star, given the mass of the star, is reasonably well known. Thus if we can evaluate the distance between the neutron star and its low-mass companion, the probability of the occurrence of x-ray eclipses can be calculated.

An evaluation of the distance between the two stars can be made. To appreciate this, let us perform a thought experiment and put the companion star at a large distance from the neutron star, so that no mass from the normal star can be gravitationally captured by the neutron star. We then move the two stars a little closer together. Eventually we will reach a point where the material on the surface of the companion star, closest to the neutron star, is attracted more strongly by the neutron star than by the gravitational field of the companion star itself. At that point, the material will flow freely from the companion star toward the neutron star. This is about the maximum required distance between the two stars. Calculations then show that for a randomly oriented binary system, with a neutron star of about 1 solar mass and a companion star of about 0.01 solar mass, the probability of observing an x-ray eclipse is about 10 percent. This probability would go up to about 20 percent and 40 percent for companion stars of about 0.1 and about 1 solar mass, respectively. Thus if all companion stars in the class 2 objects had a mass as low as about 0.01 solar mass, we would expect that out of twenty such objects, about two would show x-ray eclipses. However, none is observed. If the companion stars were all of mass 0.1 or 1 solar mass, the expected number of eclipsing systems would be even higher: about four and eight, respectively. How then do we explain the absence of eclipses in all twenty objects? Could it be that the majority of the companion stars have masses even less than 0.01 solar mass? We do not believe so. No doubt the companion stars are of low mass, but we have good reasons to believe that at least several of them have masses in the range 0.5 to 1 solar mass. What then is wrong with the results of the calculations mentioned above? Nothing, but the low-mass idea alone cannot be the whole story.

In the process of mass transfer, an accretion disk is formed surrounding the neutron star. If the accretion disk is very thick and the companion star rather small (of low mass), an observer on the neutron star might not see the companion star, for it would be obscured by the accretion disk. In fact, under those same circumstances, an observer on Earth would not see x-ray eclipses either. To see x-ray eclipses, the Earth would have to lie in or near the orbital plane of the binary system. But that would mean that the Earth also lies in the plane of the accretion disk, and those x-rays emitted in the direction of the Earth would be absorbed by the disk. Thus no x-rays would be seen on Earth. We would be able to see the x-rays only if our view direction makes a sufficiently large angle with the plane of the binary system for us to peek over the accretion disk. In other words, a conspiracy between the thickness of the accretion disk and the companion star would permanently obscure those x-ray sources that would otherwise show x-ray eclipses, and we would see x-rays only from those systems that do not eclipse. This may well be the correct explanation for why no x-ray eclipses are observed in about twenty carefully studied class 2 objects.

The low-mass companion stars in these class 2 objects should be very faint. Is it not possible to find these faint objects with powerful optical telescopes? If they were found, it would lend strong support to our idea that class 2 objects are low-mass binary systems. In fact, it is possible, although certainly not easy, and several have been found. There is a serious complication, however; the optical light from the class 2 objects comes from the accretion disk as well as from the companion star. The accretion disk light, in general, is substantially brighter than that of the star. The light from the accretion disk results largely from the bombardment of x-rays that come from the neutron star. The disk intercepts an important fraction of these x-rays, which keeps the disk at a high temperature. The exact value of the temperature depends on the x-ray flux and on the location of the disk (for example, the outer part of the disk is cooler than the inner part); the temperature could range from about 5000 to 50,000 K. At these temperatures the disk would radiate mainly optical and ultraviolet light. In many class 2 objects the x-ray flux is always so high that the light from the disk dominates that of the faint companion star. If one could only turn off the x-ray source, the light from the disk would be greatly reduced, and there might be an opportunity to observe the companion star. Nature is very kind to us this time. Class 2 objects can suddenly appear in the sky as very strong x-ray sources and remain strong for weeks or months. The x-ray flux then decreases gradually, and the sources become fainter in the months that follow; ultimately the x-rays become undetectable. They are called x-ray-transient sources, or x-ray novas. Such an x-ray-nova outburst (not to be confused with a much briefer x-ray burst) can recur with intervals of a year or so. These x-ray novas are ideal to search for the faint companion stars—after the x-ray source is "turned off."

How do we know what part of the light we see comes from the disk and what part from the companion star? If we cannot tell, we will never know whether we have actually seen the companion star. The optical spectrum from the nuclear-burning companion star shows certain characteristic features of stars. If such a spectrum is measured, a classification can be made from which, as a rule, the temperature and the mass of the star follow. The optical spectrum of the accretion disk, however, is very different. As an example, let us take the class 2 recurrent transient source Centaurus X-4, which had an x-ray-nova outburst in May 1979. The nova outburst was observed with the British satellite Ariel 5. X-ray astronomers reported their findings by means of the International Astronomical Union's telegram system. Subsequently optical astronomers pointed their powerful telescopes in the direction of the x-ray source and noticed that a previously inconspicuous starlike object had brightened substantially (figure 4). The optical spectrum was characteristic of

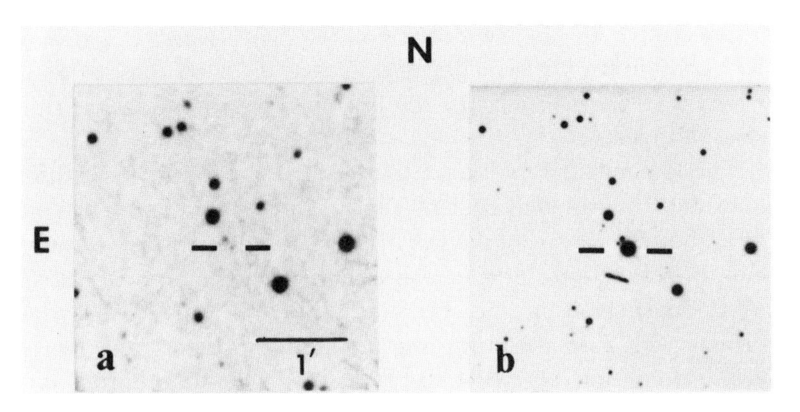

Figure 4
The optical brightening of the x-ray nova Centaurus X-4. (a) Palomar Observatory sky survey blueprint showing Centaurus X-4 in its quiescent state. (b) The discovery plate obtained by Martha H. Liller with the CTIO 4-meter telescope on May 19, 1979. The star is located between the two markers. The optical brightening is believed to be largely due to a temperature increase of the accretion disk surrounding the x-ray star. The disk is heated by the x-radiation produced in the x-ray nova outburst (not to be confused with a brief x-ray burst). By the end of June 1979, the star had faded substantially and was about as faint as shown in the photograph on the left (a). (Figure from C. Canizares, J. McClintock, and J. Grindlay, *Astrophysical Journal* 236 [1980]: L55)

emission from an accretion disk but not from a star. In the weeks that followed, the x-ray flux decreased, as did the optical flux. (During the x-ray decline, x-ray bursts were observed from this source by the Japanese x-ray satellite Hakucho.) About five weeks after the beginning of the nova outburst, the optical counterpart had become about one hundred times fainter, and the optical spectrum exhibited the features characteristic of a star of surface temperature in the range 4000 to 4600 K. This tells us that the star has a mass of about 0.7 solar mass. (Theoretical models of stars show a relationship between mass and temperature for normal stars. See the discussion by John Huchra.) In 1976 and 1977, a similar scenario had led to the observation of the low-mass companion stars of the class 2 x-ray novas A0620-00 and Aquila X-1. Interestingly Aquila X-1, just like Centaurus X-4, produced x-ray bursts shortly after the nova outburst reached its maximum; no x-ray bursts were reported from A0620-00.

To date, the masses of the companion stars of only four class

2 objects are known; two of these, Aquila X-1 and Centaurus X-4, are burst sources. They are all near 0.7 solar mass or less, which is strong support for the low-mass binary model for this class of objects.

The low-mass, close-binary model for the class 2 objects dictates that their orbital periods should range between a fraction of an hour and several days. This is a consequence of Newton's law of gravitation. The measurement of such periods clearly would be important; however, it is not easy to measure the orbital period of a nonpulsing and noneclipsing x-ray source. Nor is there much hope of observing the orbital period through the Doppler effect in the optical spectra. The emitted colors from the accretion disk are highly blurred from the disk's own rotation around the neutron star, and the colors from the companion star, if observable at all, are extremely weak. Nevertheless there are five class 2 objects for which the orbital periods have been measured with certainty. The orbital periods of four of them range from about 0.7 hour to about 19 hours, and one of them, Cygnus X-2, has an orbital period of about 9.8 days. (Its nuclear-burning companion star probably has a mass of about 0.7 solar mass). This is in good agreement with our low-mass, close-binary model. There are indications that the binary periods of Aquila X-1 and Centaurus X-4 are about 31 hours and about 8 hours, respectively; however, that is still somewhat uncertain.

Let me summarize briefly some of the general characteristics of the class 2 objects. They are binary systems that contain a very old neutron star with a nearby and also very old low-mass, nuclear-burning companion star. The mass transfer from the companion star to the neutron star is responsible for the strong, persistent x-radiation that comes from the neutron star's surface and the inner part of the accretion disk. Thermonuclear flashes on the surface of the neutron star probably are responsible for the x-ray bursts. In the presence of a strong x-ray flux, the accretion disk is heated and radiates optical light, which is substantially brighter than the optical light produced by the low-mass companion star alone.

If these global features are correct, one would expect that an increase in the x-ray flux would increase the temperature of the accretion disk and thus would increase its optical brightness. In the case of such an increase, one would expect the optical response to trail the change in the x-ray flux, since the distance from the neutron star to us on Earth is shorter than the sum of the distances from the neutron star to the disk and from the disk to us. In addition to this delay after the x-ray outburst, the optical response should be smeared out as a result of the physical extent of the accretion disk. (Different locations of the disk give rise to different delays.) The amount of delay and smearing could give us information about the geometry of the accretion disk, which cannot be obtained in any other way. Are sudden changes in the x-ray flux observed? Yes, of course; x-ray bursts. They are ideal to probe the surroundings of the neutron star. Unfortunately x-ray bursts often occur irregularly and, in general, no more than once every two to ten hours and sometimes even less often. However, with a lot of patience and coordination, the simultaneous observation of x-ray bursts with ground-based optical telescopes and x-ray observatories in Earth orbit should be possible.

In the summer of 1977, before we realized what might be learned from such observations, Jeffrey Hoffman and I organized the first coordinated worldwide burst watch. Forty-four astronomical observatories (optical, infrared, and radio) from fourteen different countries participated; the x-ray observations were made by the SAS-3 group at MIT. A total of 120 x-ray bursts were observed in thirty five days from ten different burst sources; however, no simultaneous detection was made in either the radio, optical, or infrared wavelengths.

The next year we continued our efforts with highly improved optical sensitivity. On June 2, 1978, this time in collaboration with Harvard University, we succeeded in making the first coincident detection of an optical burst and an x-ray burst from the source MXB1735-44, which has a very faint optical counterpart discovered a year earlier by Jeffrey E. McClintock. (MXB stands for MIT X-Ray Burst Source.) Since then two additional burst sources have been observed to produce such coincident events. In all cases, the optical signal trails the x-ray signal by a few seconds.

Simultaneous optical/x-ray observations are very difficult to make for the following reasons. First, very large optical telescopes are needed (preferably 90-inch diameter or larger), since the optical counterparts are so faint, but observing time on such large telescopes is difficult to get. Second, the observations have to be made at dark time (near New Moon) to reduce the sky contamination of moonlight, but dark time is precious to ground-based observers and is even more difficult to get. And finally, the sources burst irregularly and sometimes not for days at a time. It is not easy to persuade an optical astronomer to use his precious dark time on a big telescope by staring, night after night, at a star that does not burst. It takes a lot of patience indeed.

Fortunately Holger Pedersen from the European Southern Observatory has been very dedicated to these observations, and his perseverance has paid off. In the summer of 1979, he detected fifteen optical bursts from MXB1636-53, and five of these were simultaneously detected by the Japanese x-ray satellite Hakucho under the direction of Minoru Oda (figure 5). The optical flux from MXB1636-53 doubles in less than a few seconds, and it is about 3.5 seconds delayed relative to the x-ray signal. One can derive from this delay time that the radius of the accretion disk is approximately 1 million kilometers.

In the summer of 1980, observations of the same source were continued; but this time some of the optical data were taken in three spectral bands (ultraviolet, blue, and visual). This may allow us to see changes in the temperature of the disk as it is heated by the x-ray burst. We expect the disk to heat up as the x-ray burst rises and then to cool as the x-ray burst fades. The data are currently being analyzed.

Finally I would like to discuss a unique object, the Rapid Burster, which is very different from all other burst sources. At first, this object seemed a big spoiler, capable of destroying all our theories, but later it became crucially important in unraveling the burst mystery.

We discovered the Rapid Burster, whose official name is MXB1730-335, in early March 1976. We observed bursts in quick succession, as many as a few thousand per day (figure 6). Clearly this behavior was completely different from anything we had seen; the other burst sources produce typically a few to perhaps ten bursts per day. The energy in the individual bursts from the Rapid Burster varied by a factor of one hundred, in strong contrast to bursts from the other burst sources. Also the bursts from the Rapid Burster that last longer than about fifteen seconds have flat tops. After each burst, the waiting time to the next burst is approximately proportional to the total energy in the previous burst. The burst mechanism seems to involve a continuously fed storage tank of material, which can empty (causing a burst) only when it has filled up to a critical level. If

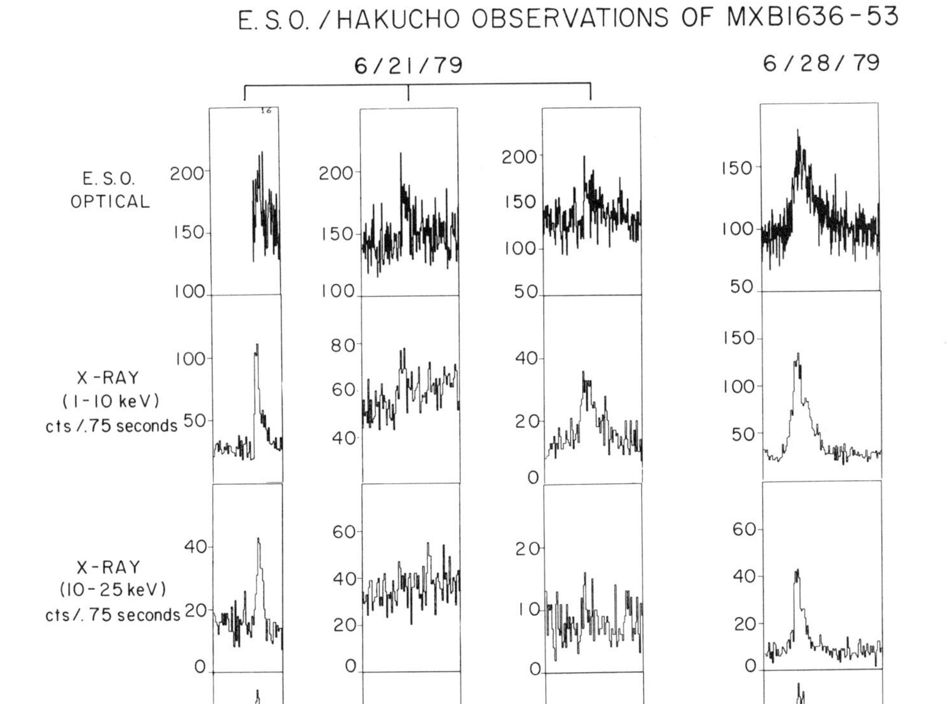

Figure 5
Four optical and x-ray bursts simultaneously observed from 4U/MXB 1636-53. The optical data are at the top. The x-ray data, obtained by the Japanese x-ray observatory Hakucho, are below the optical data in three energy bands, as indicated. The vertical scale of the optical data is intensities in relative units. The x-ray intensities are x-ray events accumulated in 0.75 second. The horizontal scale is time, indicated in hours, minutes, and seconds; each small division is separated from the next by 6 seconds. In this source, the optical response is delayed by about 3.5 seconds relative to the x-rays. This small delay is difficult to see in this figure. (Figure from Pedersen et al., in preparation, 1981)

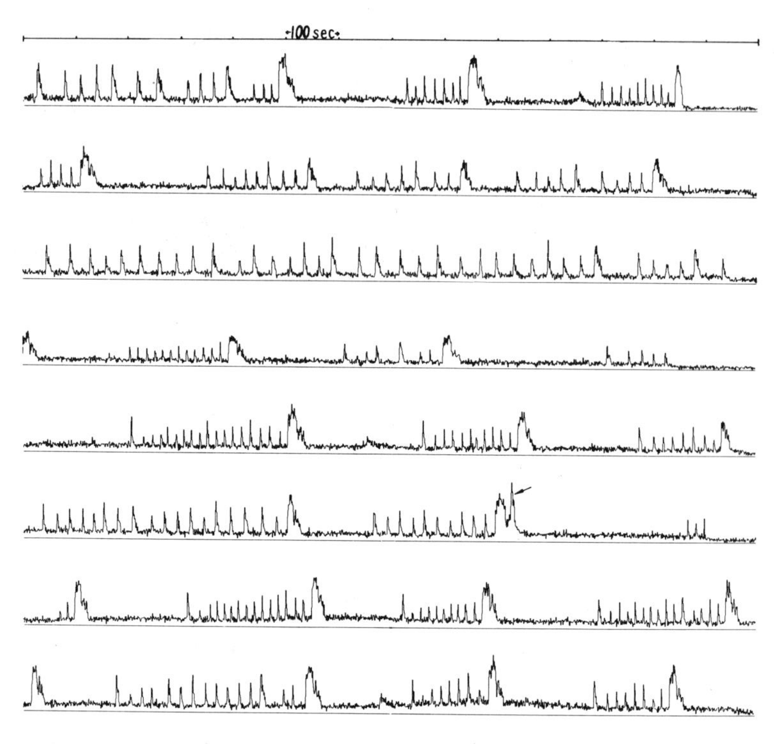

Figure 6
Discovery of rapidly repetitive x-ray bursts from the Rapid Burster (MXB 1730-335) in early March 1976. Each stretch of data is a twenty-four-minute observation made with the SAS-3 x-ray observatory. X-ray intensity is plotted on the vertical axis, and time is plotted on the horizontal axis. The time between two small tick marks at the top is one hundred seconds. Notice that for fat bursts the waiting time after a burst is longer than for skinny bursts. This is like the behavior of a relaxation oscillator. The burst marked with an arrow is not from the Rapid Burster but from the nearby source MXB 1728-34 (see figures 2 and 3). (Figure from W. H. G. Lewin, *Annals of the New York Academy of Science* 302 [1976]: 210)

a large amount of material is emptied on a given event (burst), then there is a long time until the critical level is again reached; and, if the amount emptied is small, then there is a short time to the next burst.

When we discovered the Rapid Burster we had little doubt that the bursts were due to instabilities in the accretion flow onto the x-ray star, since these bursts could not possibly be due to thermonuclear flashes. If they were, one should observe one hundred times more energy in the continuous flow of x-rays (due to accretion) than in the bursts, and such an extraordinarily high continuous flux was not observed. Thus the Rapid Burster spoiled the theory sketched above. Its repetitive bursts that came like machine-gun fire were in apparent disagreement with the nuclear flash theory that was then only in its developmental stage. (I remember very well my enthusiasm when Laura Maraschi described to me in February 1976 how thermonuclear flashes might be responsible for the x-ray bursts. Alas, when we discovered the Rapid Burster in March 1976, I lost my enthusiasm.) If nuclear flashes did produce the bursts observed from all other sources, then the spurts of x-rays from the Rapid Burster required a very different mechanism. That was not too appealing. But this is not the end of the story. Although the Rapid Burster at first looked like a spoiler, eighteen months later it became a Rosetta Stone.

The burst activity of the Rapid Burster stopped in April 1976. Luckily the source becomes burst active for several weeks approximately every half-year. In the fall of 1977, Jeffrey Hoffman, Herman L. Marshall, and I were again observing the Rapid Burster with SAS-3 when we detected, in addition to the rapidly repetitive bursts, every three or four hours a burst that looked very different. We called them special bursts (figure 7). We were certain that they came from the Rapid Burster and not from another object since the occurrence of a special burst had a noticeable effect on the pattern of the rapidly repetitive bursts. We noticed too that these special bursts were very much like the bursts from all other burst sources: first, the burst intervals were several hours; second, the burst duration and strength was about the same for all eighteen special bursts that we detected; and, third, the special bursts lasted longer at lower energy x-rays than at higher energies. This last point is very important. The spectra of bursts from all burst sources, except the rapidly repetitive bursts from the Rapid Burster, show this distinct sig-

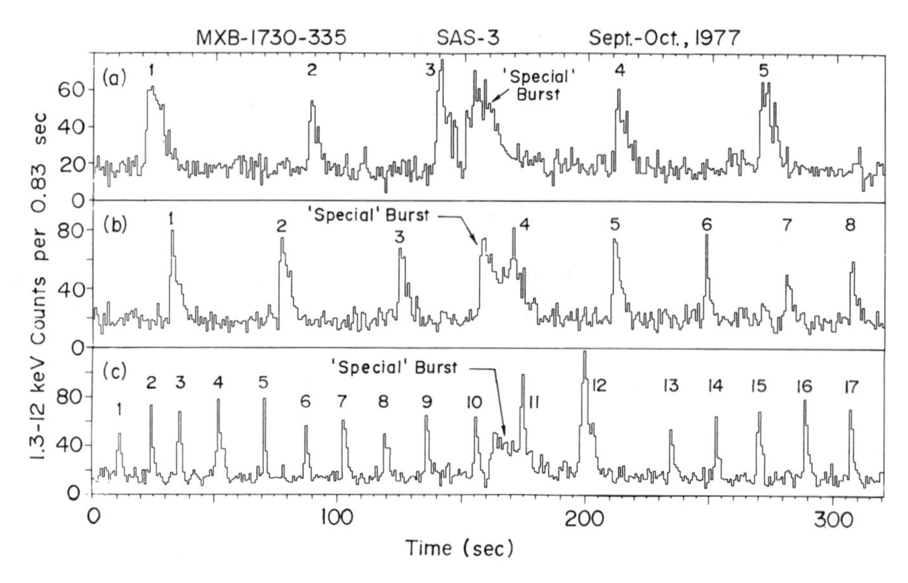

Figure 7
Discovery of special x-ray bursts from the Rapid Burster; they occur independently of the sequence of the rapidly repetitive bursts (numbered separately). Each stretch of data is about 320 seconds long (the horizontal axis is marked in seconds). The vertical axis is the x-ray intensity (1.3- to 12-keV band) in x-ray events recorded in 0.83 second. The special bursts are very different from the rapidly repetitive bursts and are believed to be the same kind of bursts (thermonuclear flashes) as the common bursts from all other burst sources. The discovery of these bursts was a turning point in our ideas of the mechanism behind the x-ray bursts. (Figure from J. A. Hoffman, H. Marshall, and W. H. G. Lewin, *Nature* 271 [1978]: 630)

nature, which is due to the cooling of a black body (the neutron star).

Now that two very different kinds of bursts were observed from the Rapid Burster, the thermonuclear flash idea came back to life. It became appealing to assume that the special bursts were due to thermonuclear flashes and the rapidly repetitive bursts due to instabilities (hiccups) in the accretion flow. If our suspicion was correct, the energy in the rapidly repetitive bursts, averaged over a day of observations, should be about one hundred times that in the special bursts. (The value of one hundred comes from the ratio of the gravitational potential energy to the nuclear fusion energy.) We measured the ratio and we found 130, which is in excellent agreement with the model. So it turned out that the special bursts were special only to the Rapid Burster, but those very same bursts were the common x-ray bursts in all other burst sources.

This discovery made in the fall of 1977 was a turning point in our thinking, and it greatly revived our enthusiasm for the thermonuclear flash model. In the spring of 1978, Joss succeeded in demonstrating, by computer simulations, that the special bursts from the Rapid Burster and the common x-ray bursts from all other sources probably have the same nuclear-flash origin.

I have come to the end of the mystery of x-ray burst sources. The broad features of these objects, and thereby the entire class 2 objects, have become clear. X-ray burst sources are collapsed objects of roughly solar mass, probably neutron stars in most cases, which are accreting matter from a low-mass, thus optically very faint, stellar companion. X-ray bursts very likely result from thermonuclear flashes in the surface layers of the neutron stars, while rapidly repetitive bursts from the Rapid Burster almost certainly result from a spasmodic accretion flow onto the neutron star.

Even though most of the mystery is solved, a number of problems are left. This is not uncommon in science; answers often lead to new questions. What is the nature of the accretion instability that produces the unusual pattern of rapidly repetitive x-ray bursts in the Rapid Burster? Why is this unusual pattern of bursts not observed in other x-ray sources? Why does the Rapid Burster become burst active at six-month intervals? How are the very old low-mass close-binary systems (the class 2 objects) formed? As proposed by George W. Clark, those class 2 objects located in globular clusters could well be produced when there is a close encounter between a neutron star and a low-mass nuclear-burning star. But how about the class 2 objects not located in globular clusters? How were they formed? The answers are not in yet. Nonetheless we have come a remarkably long way. Five years ago the class 2 objects were a complete enigma and x-ray bursts were unknown. Now we seem to be on our way to a satisfying understanding of what these objects are and how they produce the bursts. Through these bursts we have gained a new, very powerful tool to examine the properties of neutron stars, their accretion disks, and the low-mass close-binary systems in which they often occur.

There is a real danger that this new and very promising area of research will come to a halt in the 1980s. At present only the Japanese x-ray observatory, Hakucho, can make the x-ray burst observations in coordination with ground-based observers; however, its life is limited, with probably only a few more years to go. The European Space Agency may successfully launch its x-ray observatory, EXOSAT, in 1982, but that is still rather uncertain. Let us hope that the U.S. National Aeronautics and Space Administration will soon be able to approve its existing plans for the X-ray Timing Explorer (XTE). This x-ray observatory would be ideally suited for studying a wide variety of highly variable phenomena such as x-ray bursts.

Further Reading

Clark, G. W. "X-Ray Stars in Globular Clusters." *Scientific American* (October 1977).

Lewin, W. H. G. "X-Ray Burst Sources: A Mystery Solved?" *Scientific American* (May 1981).

———, and van Paradijs, J. "What Are X-Ray Bursters?" *Sky and Telescope* (May 1979).

Penrose, R. "Black Holes." *Scientific American* (May 1972).

Thorne, K. "The Search for Black Holes." *Scientific American* (December 1974).

CHAPTER 7

THE AGE AND STRUCTURE OF THE UNIVERSE

WILLIAM H. PRESS

A TALE OF TWO THEORIES

At present we know of only three ways to estimate cosmic ages. The first way is by radioactive dating of minerals on the Earth or materials from elsewhere in the solar system (such as the Moon and meteorites). This beautiful and precise work is described in part in this volume, in chapter 3 by Whipple and Franklin. The result, of course, does not determine the age of the universe but only the age of the Earth (or of the solar system). This leaves completely unanswered the question of whether the Earth was formed very early on, so that its age is about the same as that of the universe, or whether it was formed rather recently on a cosmic time scale, with the universe being much more ancient. In fact, the currently accepted view is that the Earth is certainly no older than about half the age of the universe and possibily only about a quarter as old.

The second way to get some idea of a cosmic age is by observing the current state of evolution of stars in globular clusters. By theoretical modeling of the evolution of stars in burning their nuclear fuel, it is possible to predict how a population of stars, containing a variety of different masses, should look—what their distribution of colors and brightnesses should be—when they have lived a certain age. One then works backward to deduce from observations a consistent age estimate for a particular cluster. John Huchra discusses this technique in some detail. But here we have the same problem as with the Earth: we do not know for sure that the globular clusters are old objects, cosmologically speaking. In this case, however, the currently accepted view is that the clusters are old, almost as old as the universe.

The third way tells in principle the age of the universe directly, by observing its intrinsic internal dynamics. The universe is expanding; galaxies are all rushing away from each other. There is no edge to this expansion, no external void into which the expansion is taking place. Rather, as Eddington emphasized some decades ago, one should think of the volume surrounding every galaxy as increasing, so that its nearest neighbor galaxy becomes farther and farther away as time goes on. If we graph the average volume surrounding each galaxy and extrapolate back in time, we find that the volume was zero a finite time ago. It is meaningless to extrapolate back any farther than that time. We call the time when the galaxies were compressed to zero volume the beginning of the universe, and we call the start of its subsequent expansion "the big bang."

The overall expansion of the universe is very slow by human standards. In some cases, expansion velocities may be large, but the distances to be covered are even larger. During my lifetime, the volume of the universe (or the volume per galaxy, a slightly less mind-boggling concept) will expand by less than three parts in a hundred million. It is not surprising, therefore, that the structure of the universe appears static to casual observation. In a human lifetime, we see a mere snapshot of its actual evolution.

The universe is filled with galaxies. Our own galaxy is called the Galaxy or the Milky Way. Our nearest large neighbor is Messier 31, usually called the Andromeda Galaxy because it lies in the constellation Andromeda. If we look on a larger scale, say a few million light-years, we find that we and the Andromeda Galaxy are the biggest members of a local group of galaxies. Groups of galaxies typically have a few or a dozen members, and our group is not untypical. Small groups of galaxies, such as our local group, are often found to be clustered in the vicinity of much larger, much richer clusters. For example, we are on the outer fringes of a region dominated by the great Virgo Cluster of galaxies. The center of the Virgo Cluster is roughly some 50 million light-years distant. The biggest cluster in our neighborhood of the universe is the Coma Cluster, which contains several thousands of galaxies.

As we look farther into the universe, we always see the same pattern: galaxies collected into groups, those groups into larger groups, and so on. But as we get to larger and larger scales the structure of the universe starts looking smoother and smoother. The clustering does not stand out in high relief as it does on smaller scales. If we look out to the largest scales accessible from the Earth-based observations, we see a picture like that shown in figure 1. The picture shows the entire northern hemisphere of our sky, with the positions of well over half a million galaxies plotted. The data for figure 1 were gathered over many years by the astronomers Shane and Wirtanen, and the computer-

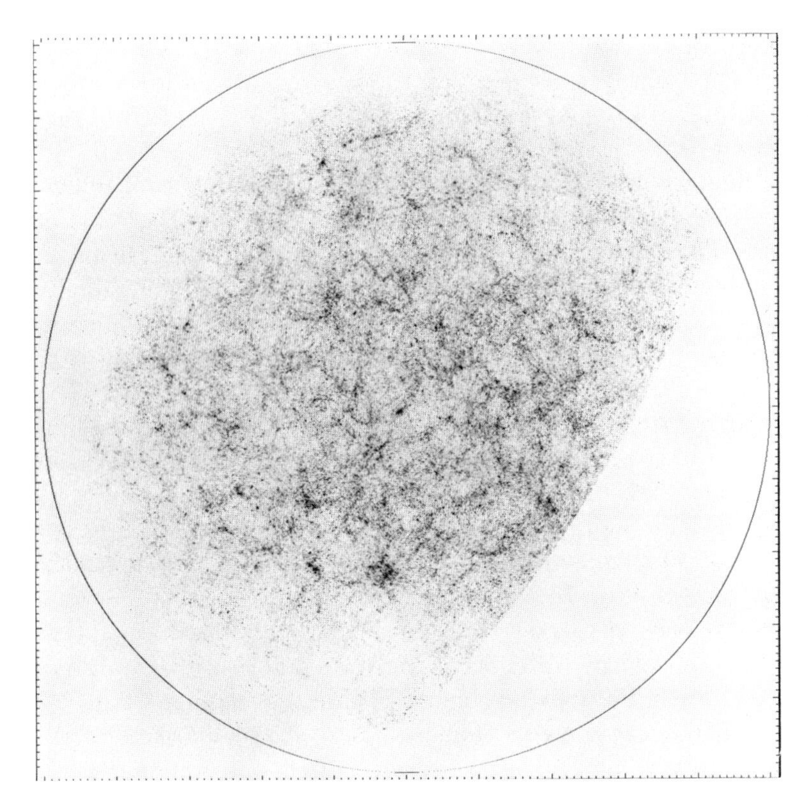

Figure 1
Galaxies in the northern hemisphere of the sky. Each dot is one to ten galaxies, with the lightest dots representing one galaxy and the darkest ten galaxies. (Reproduced with the permission of P. J. E. Peebles of Princeton University)

generated image shown was produced and interpreted by the cosmologist James Peebles of Princeton University.

Each dot in figure 1 represents not only galaxy but a number of them (typically three to ten). The brighter the dot, the more galaxies it represents. We can conclude from the picture that although there seems to be some structure on all scales, on the very largest scales, the distribution of galaxies is remarkably homogeneous (the same in all places) and isotropic (the same in all directions).

In 1924 Edwin P. Hubble, working at the Mount Wilson 100-inch telescope, established conclusively that the spiral nebulas (what we now call galaxies) were in fact located at distances measured in millions, or many millions, of light-years from

Earth. Over the next five years, Hubble's observations showed that the galaxies were not, on the average, at rest with respect to the Earth but were expanding away from us and from each other.

Suppose there is a galaxy located at a distance of 1 million light-years from us and another located beyond it (in the same direction) an additional 1 million light-years distant. The closer of these two galaxies might be observed to be receding from us at a velocity of, say, 20 kilometers per second. An intelligent being on a planet of a star in that galaxy sees us receding from him, in the opposite direction of course, as our two galaxies fly apart. Since all the galaxies are receding from each other, he sees the second galaxy as going away from him with a velocity of about 20 kilometers per second. Now jump back to our point of view: the more distant galaxy must be receding from us at a rate of 40 kilometers per second ($40 = 20 + 20$) so as to make the picture a consistent one. This kinematic fact—that the recession velocity of galaxies is proportional to their distance (at least until the velocity starts becoming comparable to that universal speed limit, the speed of light)—is known as Hubble's Law.

Hubble was the consummate observer. On the basis of the incontrovertible evidence of careful, direct observation, he first recognized that the universe is expanding. In this series of papers dealing with the interaction of theory and observation, he must be counted as a great hero. To my chagrin (since I am a theorist), I must now relate the rather interesting story of how two great cosmology theorists (who arguably are the two greatest theoretical physicists of all time) came off a good deal less well than Hubble in their theoretical work on the same question: the large-scale dynamics of the universe.

It is an observational question to ask, When was the expansion of the universe first discovered? It is a theoretical question, on the other hand, to ask, When was the theory of gravity sufficiently advanced that the expansion of the universe should have been predicted? My thesis here is that it ought to have been predicted not once but twice: first by Isaac Newton in the year 1692, on the basis of his theory of universal gravitation; and second by Albert Einstein in 1917, when he ought to have noted that his new theory of gravitation, General Relativity, while modifying Newtonian gravitation in many respects, actually strengthens and makes more precise Newton's 1692 (non)prediction that the universe cannot be static.

Newton could well have deduced that the universe was expanding. It is rather interesting that he failed to do so. We know that he thought about exactly the relevant questions, from a letter written on December 10, 1692, to Dr. Richard Bentley. The letter reads in part: "If the matter of our sun and planets and all the matter of the universe were evenly scattered throughout all the heavens, and every particle had an innate gravity toward all the rest, and the whole space [volume] throughout which this matter was scattered was but *finite,* the matter on the outside of this space would, by its own gravity, tend toward all the matter on the middle of the whole space and there compose one great spherical mass." To rephrase in modern language, if any finite amount of matter is put, at rest, into a finite region of an otherwise empty space, then it will fall in on itself because the force of gravitation is always attractive. Newton was correct on this point. But then he continued. "But if the matter was evenly disposed throughout the *infinite* space, it could never convene into one mass; but some of it would convene into one great mass and some into another, so as to make an infinite number of great masses scattered at great distances from one to another throughout all that infinite space. And thus might the sun and fixed stars be formed."

The sense is clear enough. Newton believed that an infinite universe of constant mass density could collapse locally, but that once such local collapses had occurred, one would be left with, on the large scale, a static equilibrium of stationary, or perhaps randomly drifting, bodies. This is wrong. Einstein, with the added subtlety of his new theory of gravity, understood the error, and it took him a decade (and, more importantly, Hubble's observations) to reconcile him to the correct answer. But let us try to understand Newton's error in terms that he himself could well have posed.

I have used my computer to generate snapshots of one region (bounded by a square window that frames each snapshot) of a Newtonian cosmology. Following Newton's idea of local collapse into great masses, I have instructed the computer to collapse the initially smooth mass density into letters of the alphabet (figure 2) rather than galaxies. To Newton, each letter might have represented a star; today we would rather imagine each letter as a galaxy. (I call these alphabet-soup cosmologies.)

This Newtonian cosmology extends far beyond the boundaries of the snapshot—in fact (in Newton's conception) infinitely

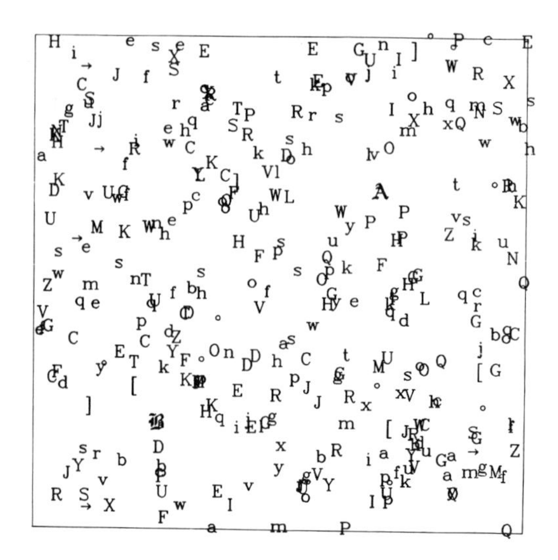

Figure 2
A portion of the alphabet-soup cosmology, which extends to infinity. Each letter represents a galaxy.

far. Also among all the Roman letters in the figure are two Gothic letters, A and B. There is nothing special about these galaxies, except that I want to refer to them conveniently in the discussion.

Let us consider the first part of Newton's discussion, concerning a universe in which matter occupies only a finite region within an empty space (figure 3). The finite region chosen happens to surround A, and it contains galaxies with precisely the same alphabetic labels and precisely the same arrangement as in figure 2. Newton explained, correctly, that the dynamics of this system, under the influence of gravitation, is all to fall in onto the center. I show this in figure 4, which is the same cosmology as that of figure 3, but just a bit later in time.

Looking back at figure 2, let us now turn to the second part of Newton's argument, about an infinite universe. Newton thought that if the mass were distributed evenly (not just in the square but all the way to infinity outside its boundaries), then it could not all fall in on any center because there is no particular center that is preferred. The universe cannot fall in on A because it is also equally attracted toward B, and so on. This is what one now calls, in physics, a symmetry argument. On the surface it is a very convincing argument. Newton was not stupid for making it; he was wrong, however.

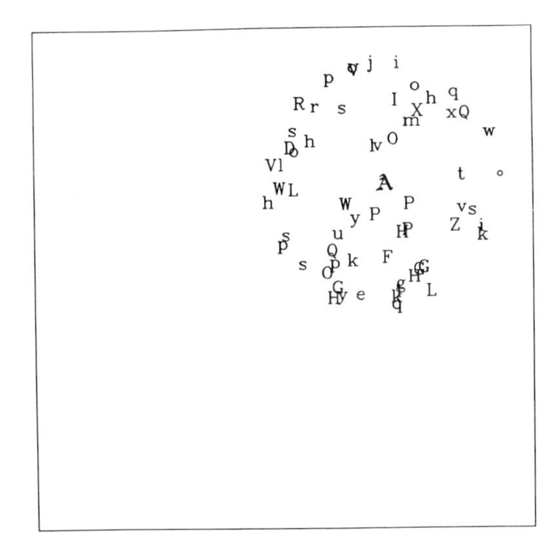

Figure 3
A universe in which matter occupies only a finite region of space, surrounding the Gothic letter A. The arrangement of galaxies around Gothic A is the same as in the infinitely filled universe of figure 2.

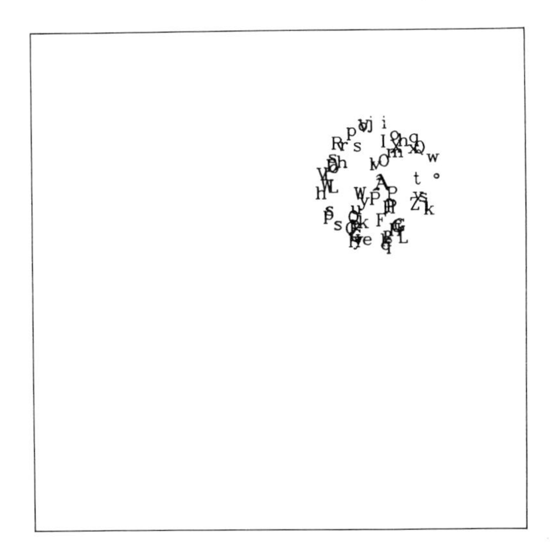

Figure 4
The same universe as in figure 3 but later in time.

Newton's error follows from a theorem in classical potential theory, which we today call, ironically, Newton's Theorem (actually one of many theorems devised by this genius). The subject of the theorem is the gravitational force that would act on an object that is inside an empty space, for example, hollowed out in an otherwise uniform density of matter. The special case that concerns us is that of a perfectly spherical hollow cavity surrounded isotropically by matter on the outside. The theorem then says simply that there is no gravitational force at all inside the cavity. An object put anywhere inside will be attracted neither to the center nor to the edge. Any two objects put inside will orbit each other exactly as if they were in a completely empty space, in a perfect Keplerian ellipse, and so on. This theorem is completely rigorous, and it turns out to be rigorously true in Einstein's Theory of General Relativity (where it is called Birkhoff's Theorem).

Let us now see what is the consequence for our alphabet-soup cosmologies. Imagine hollowing out a spherical cavity, where A used to be, within which Newton's theorem will apply. We could now put any gravitating system into that cavity. Let us choose to be perverse and put back exactly the same collection of objects that was just removed—the masses surrounding A (shown in figure 3). A short time later, according to the theorem, these masses must have collapsed just as they did in figure 4. Therefore if we evolve figure 2 in time, we ought to reach a state of affairs shown in figure 5. Or should we?

Figure 5 cannot be correct because there was nothing special about A and the objects in its vicinity. We could have done the same process around B. Then, according to Newton's theorem, the letters around B would also have had to collapse onto B. There seems to be a contradiction here, almost a reductio ad absurdum of the theory. No matter where one chooses a center and no matter what radius sphere one draws, Newtonian gravity predicts the collapse of its contents.

In fact, there is one simple, noncontradictory way out of the apparent paradox. Although Newton failed to see it, it is quite simple. Everything collapses onto everything, uniformly. The correct snapshot to follow figure 2 in time is not figure 5 but rather figure 6. In figure 6, the pattern of letters around A is just the same and has just about the same scale as figure 5. But in figure 6, the letters around B have also become closer. Since the letters around all points between A and B have become

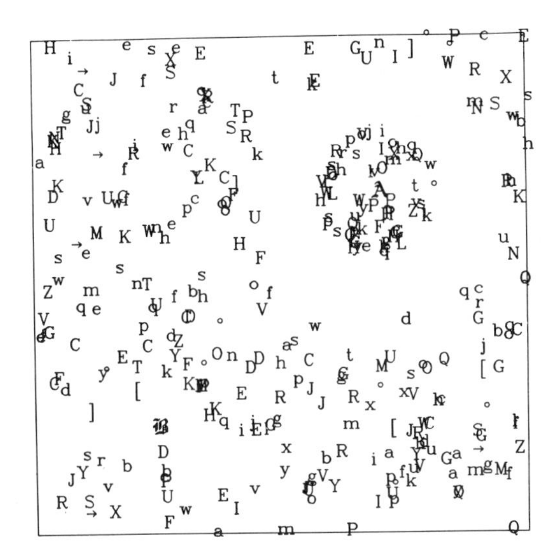

Figure 5
The alphabet-soup cosmology of figure 2, in which the galaxies around Gothic A have collapsed toward A, but all other galaxies have remained fixed. This is actually an incorrect evolution of the cosmology in figure 2.

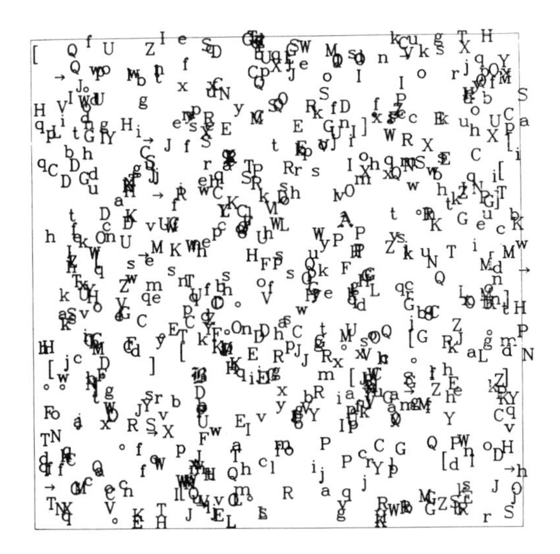

Figure 6
A time evolution of the cosmology of figure 2, in which all galaxies have moved closer together.

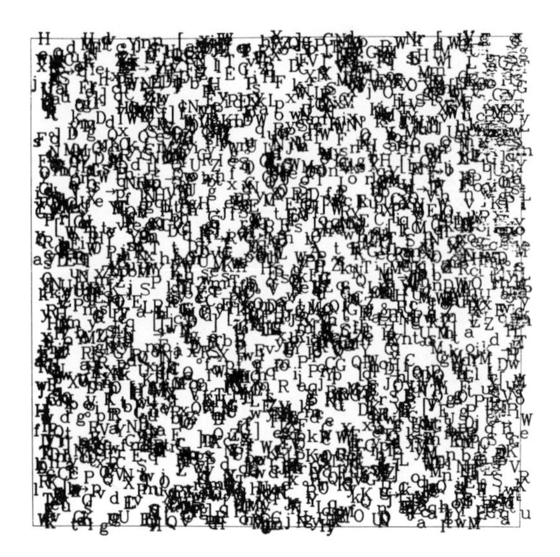

Figure 7
A time evolution of the cosmology of figure 6, in which all galaxies have moved even closer together.

more compact, it follows that A and B must have moved closer toward one another, as the figure shows. Remember that the cosmology extends much farther than the window shown. In figure 6, some letters not previously visible have moved into the window.

Figure 7 shows the alphabet-soup cosmology even later on. Looking closely, one can find the Gothic letters A and B and the same pattern of other letters around them. Figure 7 shows, in fact, a late stage in a collapsing universe.

Newton could and should have realized that as a consequence of his theory of universal gravitation, the universe was collapsing. Had he done so, he might have written to Dr. Bentley that only the vastness of the cosmic distance scale prevents the collapse from being apparent to the human observer.

How does it come about, then, that we live in an expanding, not a collapsing, universe (although it may yet collapse at some time in the distant future)? This is answered by a simple yet deep and fundamental fact about the laws of physics. For every dynamical evolution that obeys physical law, in particular the laws of gravity, there is another possible evolution that looks like the first one run backward in time. If a collapsing Newtonian cosmology is possible, then an expanding one is also possible. This would be simply achieved by imparting to the particles an initial outward velocity sufficient to counteract their gravitational attraction. It is all a question of initial conditions.

Newton himself had some understanding of this principle of time-reversal symmetry, as it is now known. He might have mentioned to Dr. Bentley that although a collapsing universe seemed more natural, starting as it does from initial conditions of uniform density matter at rest, an expanding universe, in which every mass increases its distance from every other mass uniformly, could also be possible. Nature, of course, avails itself of this latter possibility. The correct order of events in our universe is figure 7, then figure 6, then figure 2. Some further stages in the evolution will be shown below.

One might well ask whether in all the time between Newton and Einstein anyone recognized Newton's error. I believe the answer to be that the theoretical error was not recognized, but it is known that there was some recognition that Newton's picture of an unevolving cosmos led to observational contradictions.

Newton's own close friend, Edmund Halley, first noted a major difficulty. (Halley is best known today for his identification of the periodicity of Halley's comet; his greater contribution, usually forgotten, is that he encouraged Newton to write up his scientific results in gravity and mechanics. Newton's great *Principia* was published at Halley's own expense in 1687.) In 1720 Halley deduced that an unchanging cosmos would gradually fill up with the light produced by stars so that the celestial sphere should appear luminous instead of dark.

This argument, which is essentially correct, was rediscovered by P. L. de Cheseaux in 1744 and again, almost a century later in 1826, by Heinrich Wilhelm Olbers. Perhaps unfairly, it is generally referred to as Olbers's Paradox.

The resolution of Olbers's Paradox is quite simple: the sky is not bright because it has not had time to fill with light. It has not had time because the universe began, in the big bang, only a finite time ago. (A contributory factor, though less important, to the resolution of Olbers's Paradox is that in an expanding universe, light is red shifted and made less intense by the expansion of the universe.)

Matters finally did come to a head with Einstein's development of the Theory of General Relativity. General Relativity is a theory of gravitation. It attempts to explain the same range of physical phenomenology as does Newton's theory of universal gravitation, but its mathematical formulation and some of its predictions are a bit different from Newton's theory. For example, the force law between two bodies—proportional to the inverse square of the distance between them in Newton's theory—is slightly different in General Relativity. A common feature of Einstein's and Newton's theories is that the force of gravity is completely attractive, never repulsive. Things always fall toward each other (unless, of course, they have been initially flung apart). In General Relativity, exactly the same paradox that was represented in figures 2 through 5 appears. Further, because of the relative mathematical precision of the theory, that paradox comes forward in a much more forceful way than before. It is impossible to write down a consistent set of equations in Einstein's theory without its smacking one in the face that the universe has to be either expanding or contracting.

Einstein recognized this fact within the first year or so after he wrote his theory, in about 1917. This was twelve years before Hubble's announcement; Einstein could not bring himself to believe his theory's own straightforward prediction. He therefore added an extra term to the equations of the theory, called the cosmological constant term. This term has no basis in experiment whatsoever. Its only purpose is to allow a static universe: it adds a repulsive part to gravitation (I intend the pun) so that the matter in a static universe can repel itself just enough to avoid collapse. Later, after Hubble's discovery, Einstein offered the opinion that the cosmological constant was "the biggest blunder" of his life.

Like Newton, Einstein could have predicted the expanding universe. Like Newton, he had invented a theory more powerful than his own confidence in it.

I want to return now to one important issue. Many people who are not cosmologists have the mistaken view that there is an edge to the big bang as it expands and that outside that edge is empty space. I have tried to make clear in the illustrations that this is not the case. The pictures themselves have an edge, but if I had had the computer draw the boundaries a bit larger, one would have seen further regions of alphabet soup. As a theoretical model, the universe can be extended arbitrarily far, so that it can be infinitely large.

Infinities are concepts that science has difficulty wrestling with. They are rightly suspect in any theory. Infinities can (and do) occur in many theoretical models of our physical world. But it is important that they should not occur in any of the derived, measurable predictions of those models. We want to see now how a cosmology of infinite theoretical extent is rendered, in practice, observationally finite.

The key point is the same one that resolved Olbers's Paradox: the universe is of finite age. Therefore light can have traversed only a finite distance since the beginning of the universe. It is impossible for us now to know, even in principle, anything about the universe outside of this finite distance. It is the observational edge of our universe and is called our cosmological horizon.

When we say that the universe is completely smooth and without an edge, we really mean that it is completely smooth as far out as we could possibly have seen by now and that there is no physical edge but only a natural current limit on our observations. As time goes on, the limit of our observations creeps outward. Theoretically it is possible that as time goes on, we shall begin to see an actual physical edge of the matter in the big bang, with empty space beyond it. Our theoretical model does not require this, however; it is able to extend the smooth universe infinitely far. And there is not the slightest observational evidence to gainsay this model.

Figures 8 through 10 show again (now in the correct time ordering) the evolution of the alphabet-soup cosmology. Now added to the figures are shaded regions around the galaxies A and B. (At a time preceding that of figure 8, the situation would appear as in figure 7, with shaded areas so small as to include only a single letter.) These regions show the distance out to which signals from A or B have had time to travel since the beginning of the universe. Equivalently they show the region within which signals have had time to travel to A or B—in other words, A or B's cosmological horizon.

There are two important points to note. The first is that the cosmological horizons grow in absolute size (radius) with time. The second is that the horizons grow to include more galaxies with time. This second point does not follow from the first one, since as time goes on the space between the galaxies also increases. In other words there is a race between the expansion of the universe and the expansion of the cosmological horizon. The important point is that the cosmological horizon always wins this race. The cosmological horizon starts off including only a very small (or even zero) number of galaxies. But as time goes on, it expands to include more and more of the universe.

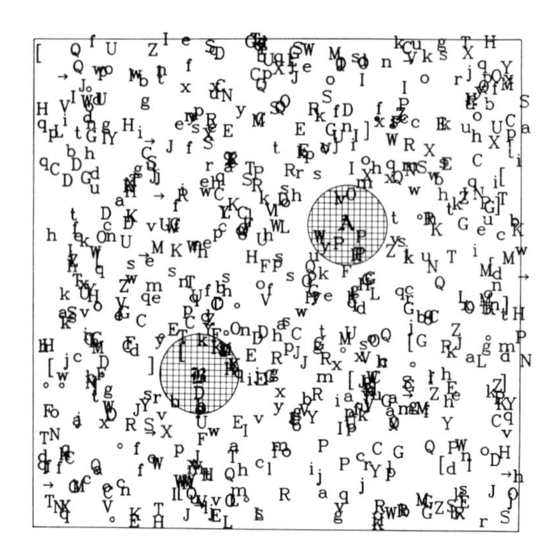

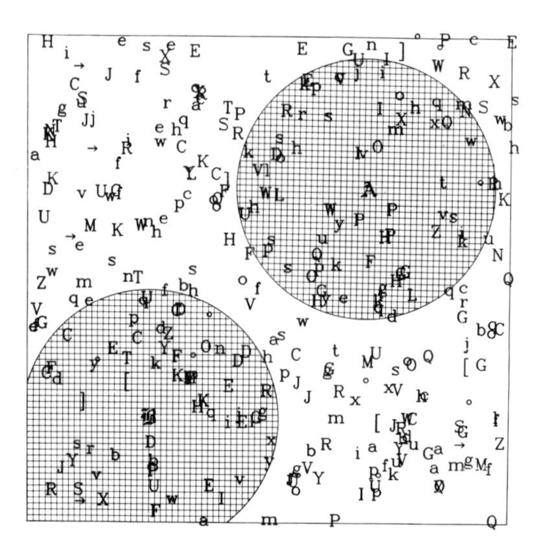

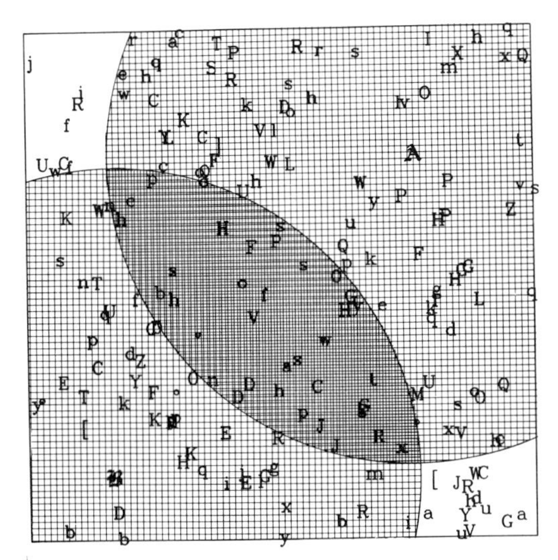

Figures 8–10
An expanding, evolutionary sequence of the alphabet-soup cosmology. Shaded regions around the galaxies labeled as Gothic A and B indicate the distance out to which A or B could have communicated since the beginning of the universe.

Thus although the universe began at infinite density, it did not begin with complete communication across its extent. As we go backward in a cosmology, the cosmological horizon includes fewer and fewer galaxies, even though these are closer and closer together. The initial state is one of infinite density but total communicative (causal) disconnection.

In figure 10, A and B still are not able to see each other. A's horizon does not include B, and vice-versa. It makes no difference that their horizons overlap. At the instant shown, A and B reside in separate universes, for purposes of communication.

One feature of General Relativity is completely novel and has no analog in a Newtonian context: matter produces not only a gravitational field but also an actual geometrical curvature in space. (In fact, in General Relativity, these two effects are mathematically inseparable.)

It is quite difficult to visualize the meaning of geometrical curvature of a three-dimensional space, so it is useful to think instead of a two-dimensional analog. Space is like the surface of a horizontal rubber sheet, with elastic properties similar to, though not quite the same as, an actual elastomer. Some sand sprinkled on the rubber sheet in one spot causes the sheet to bulge downward into a stretched, bowl shape. If we add more sand, the bowl gets deeper and the rubber more stretched.

If we keep adding sand, at some point an actual rubber sheet will tear. There is a similar effect in the space geometry of General Relativity. If the density of matter in some volume exceeds a critical value (which depends on the size of the volume and the rate of expansion of the matter inside it), then the bowl shape actually pinches off into a sphere and detaches itself from the rest of the rubber sheet. It is a little bit like a drop of water forming on a moist ceiling and finally detaching itself. Instead of residing on an infinite rubber sheet, the sand now finds itself inside the surface of a topologically closed sphere.

If the mass of the individual letters in the alphabet-soup cosmology is sufficiently large, then the square window plotted in the figures will represent not one part of an infinite plane but one part of the surface of a very large sphere (much larger than the window). In this case, if I extend the window to the right so that it becomes a long strip, my computer eventually will start drawing exactly the same pattern of letters that occurs at the left-hand edge. These will not represent new galaxies but rather the same galaxies as before. The strip will have closed on itself; it should be pasted together into a large ring that represents an equatorial band around a closed universe. At present we do not know for certain whether our universe is topologically open or closed. We do know that if it is closed, its radius is much larger than our present cosmological horizon or window. We cannot observe its closure directly, therefore, but can hope only to deduce the fact of its closure by measuring the density of matter that we see and comparing that density to the breaking strength of space as it is given by Einstein's theory.

In conclusion, theory is sometimes more powerful than it is admitted to be by the very theorists who invent it. Newton's theory could have predicted the expanding universe, but he missed seeing this prediction. Einstein's theory did predict the expanding universe, and Einstein refused to accept that the prediction was meaningful until the observational evidence forced him to do so.

These statements of fact should not be regarded as criticism of either of these two giants. The ultimate test of reality is always observation. Wrong theories can make predictions too. Restraint on the part of theorists is often a virtue. I do draw one conclusion, however: the ability of our species to represent models of physical reality that go beyond all of the observational knowledge of the moment is awesome and startling. I do not think that we have a good understanding of the nature of this ability or of its long-term consequences to our society.

Further Reading

Author's Note: My intent in this paper has been to explain scientific concepts rather than to present a serious historical study. I have drawn heavily on the excellent historical and bibliographical material in C W. Misner, K. Thorne, and J. A. Wheeler, *Gravitation* (San Francisco: Freeman, 1973). The reader should refer to this book, especially Box 27.7, for references to both the scientific and historical literature.

See also:

Gott, J. R., Gunn, J. E., Schramm, D. N., and Tinsley, B. M. "Will The Universe Expand Forever?" *Scientific American* (March 1976).

Harrison, E. R. "The Paradox of the Dark Night Sky." *Mercury* (July/August 1980).

JOHN HUCHRA

THE COSMIC CALENDAR

The simplest goal of every observational cosmologist is to measure the age of the universe—to prove that he is right and to predict the eventual fate of the universe. Unfortunately the universe is a bit vain and is unwilling to reveal this secret. Astronomers are persistent, however, and I will try here to tell you a bit about the history of our search.

When cosmologists talk of cosmic ages, three time scales are commonly cited: the age of the Earth, the age of the oldest stars, and the age of the universe. These time scales are listed in this order because it is hoped that their estimated ages will be in the same order, although this has not always been the case.

The age of the Earth is most important as the lower limit to the age of the universe. By definition, the whole shooting match must be older than the Earth or even the solar system.

Not counting Bishop Ussher's seventeenth-century estimate of the Earth's creation date as 4004 B.C., the first serious estimates of the Earth's age were made by geologists in the early part of the nineteenth century. They used estimates of the rate of deposition of sedimentary rock to derive ages of a few hundred million years. In the late nineteenth century, the physicist Lord Kelvin challenged this age. He calculated the cooling time of the Earth from a totally molten state to its present condition (cool crust, molten core) and found an age of only 40 million years. This discrepancy was later resolved by the discovery of radioactivity in rocks, for this internal power supply could explain how the core of the Earth remained molten.

The discovery of radioactivity also provided the clue to the currently used technique for measuring the age of the Earth and elements: radioisotope dating. The earliest estimate using this technique gave an age of 1.8 billion years in 1917. The technique was refined in 1956 to give an age of 4.6 billion years. The accepted estimate of the Earth's age has not changed significantly since then. Unfortunately this is the only one of the three time scales in which we have any degree of confidence.

Figure 1
The globular cluster NGC 7078 (M 15). This is at a distance of approximately 40,000 light-years. (Photograph by Hale Obsevatory)

The second time scale is the age of the oldest stars. The oldest stars astronomers have yet found are most easily studied in globular clusters (figure 1). Globular clusters are very dense aggregates of stars that are thought to have been formed at the same time or even slightly before galaxies themselves formed—within the first 10 percent of the age of the universe as a whole. Relative to these old-timers, our Sun is the new kid in town. (Most of what I will say about the age of the universe is in terms of the most widely accepted theory of the universe, the big-bang model, in which the universe does have a beginning and thus a definable age.)

We can determine the ages of the globular clusters and the stars within them because stars evolve. Their brightness (luminosity) and color (surface temperature) change with time in what theoreticians think is a reasonably predictable manner (figure 2). Massive stars evolve faster than less massive stars; they have hotter central temperatures and burn their nuclear fuel (hydrogen) faster. Because all the stars in these clusters are thought to have formed at the same time, the distributions of

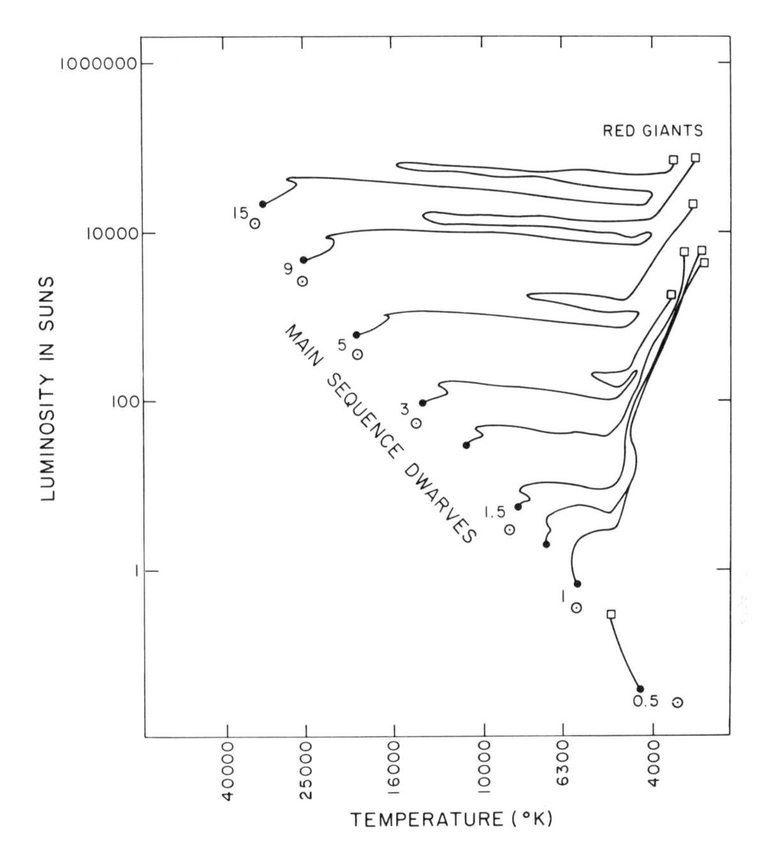

Figure 2
Evolutionary tracks for stars of masses 1 to 15 solar masses (M⊙) plotted as luminosity in solar units (L⊙) versus surface temperature in degrees Kelvin. Each track is followed by a star of the indicated mass. The end of each track, indicated by a box, marks the end-point of stellar evolution. A star of 1 M⊙ takes 10 billion years to go from a main sequence dwarf to a red giant, while a star of 15 M⊙ takes only 12 million years to evolve from dwarf to giant. (Adapted from work by I. Iben)

the theoretical models of stellar brightness and color with time can be compared to the observationally measured distributions of stars (figures 3 and 4). The observational distribution is called a Hertzsprung-Russell diagram after a well-defined region of the luminosity-color diagram, and the theoretical distributions are called isochrones, meaning lines of the same age.

As is usual in astronomy, this technique is not as simple as it sounds. The theoreticians must include guesses about certain properties of the stars, for example, their atomic parameters and chemical composition. The observers cannot measure exactly the quantities theoreticians predict (visual luminosity versus total luminosity or color versus temperature) and must struggle to derive a conversion between these systems and measure accurate distances to the clusters. Nonetheless theoreticians and observers have put forward estimates for stellar ages that have ranged from 4 billion years in the late 1940s up to a high of 26 billion years in the early 1960s. Estimates of this time scale have now settled down to between 8 billion and 18 billion years, based primarily on theoretical work by P. Demarque and observational work by A. Sandage and collaborators.

The last of our time scales, the age of the universe itself, can be derived from the expansion rate of the universe. In the big-bang model, the universe expands from an initial singularity—a time when the universe was very dense and hot and when the laws of physics as we know them are difficult to apply—so that the speed of any object relative to any other object is just proportional to the distance between them. (For a more detailed account of the very early universe see *The First Three Minutes* by Steven Weinberg.) Three common analogies for this expansion are dots on the surface of an expanding balloon, or raisins in a rising bread pudding or, more crudely (because it has a center, while the universe does not) an explosion. And the simple dynamical explanation is: the faster things move away from you, the farther they have gotten. This leads logically to a direct proportionality between velocity and distance for objects, which is also a measure of the expansion rate. This expansion rate is quoted as the ratio of velocity to distance: miles per hour per mile, or, as astronomers would say, kilometers per second per megaparsec. (A megaparsec is approximately 3 million light-years, or 20 billion billion miles.) This ratio also has units of reciprocal time; thus the reciprocal of the ratio provides an estimate of the age of the universe, or the time it has taken

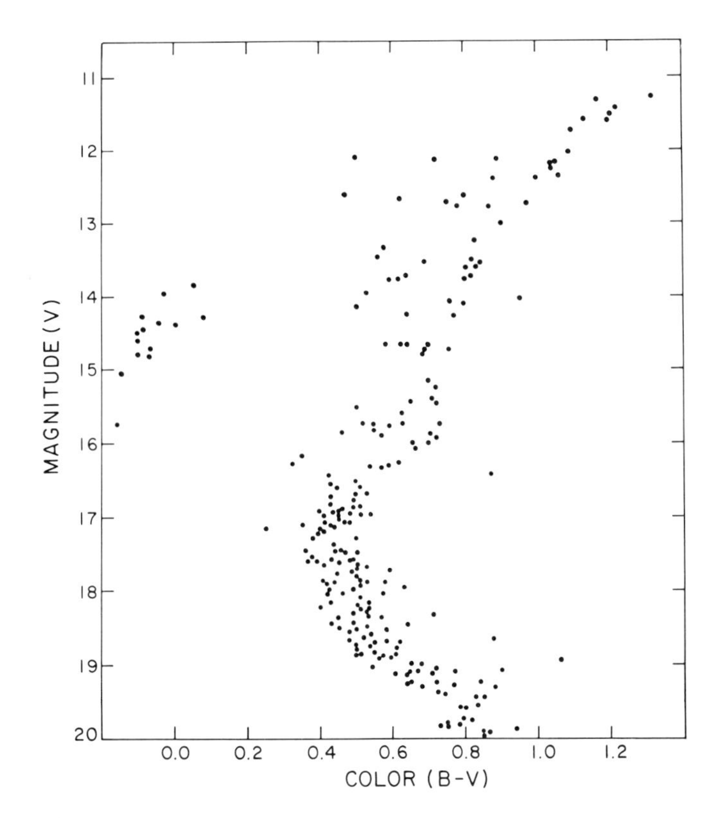

Figure 3
The observed distribution in visible luminosity versus color for some stars in the globular cluster NGC 6752. Luminosity is measured by *V magnitudes,* with higher values corresponding to less bright stars; color is measured by *B-V magnitudes,* with high values corresponding to redder stars and lower values corresponding to bluer stars. (The magnitude scale is a measure of brightness. A *difference* of one magnitude corresponds to a *ratio* of brightness of about 2.5, with brighter objects having smaller magnitudes, by convention. The V magnitude is the magnitude at green-yellow wavelengths; the B magnitude is the magnitude at blue wavelengths; and the B-V magnitude is the difference of these two magnitudes.)

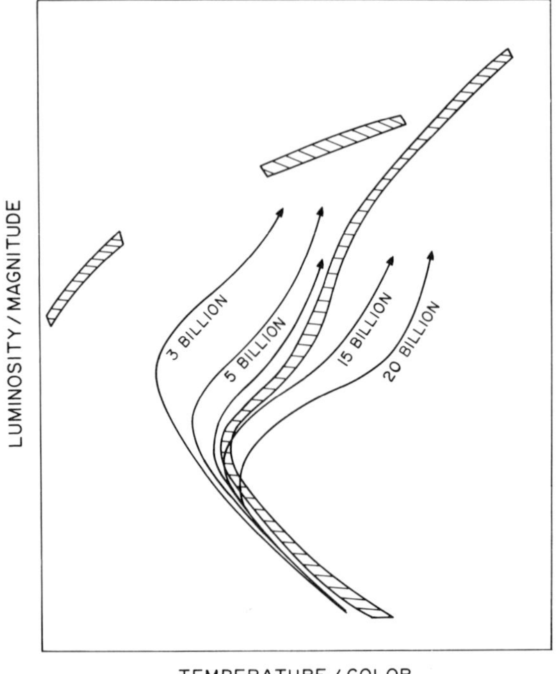

Figure 4
The curve representing the distributions of stellar luminosity and temperature observed in a star cluster (broad lines) overlaid on theoretical distributions of different ages.

galaxies to reach their present separations with their present velocities.

The first calibration of this expansion rate, the velocity-distance relation, was made by Edwin Hubble in 1929. Using galaxies as test particles, he measured their radial velocities (velocity along the line of sight) by the Doppler shifts in their spectra. (See chapter 5, Harvey Tananbaum, for a discussion of Doppler shifts and their effect on the spectra of moving objects.) Hubble also tried to measure the distances of these galaxies by looking for their brightest stars in order to compare them to the brightest stars in our own galaxy. Because apparent luminosity is proportional to inverse distance squared (a candle placed twice as far away as another candle of the same true brightness would appear only one-quarter as bright), if he knew the true distances and thus the luminosities of the stars in our galaxy he could estimate the distances of the other galaxies by assuming their brightest stars are like our stars. Because of Hubble's pioneering work, this ratio between velocity and distance as measured at the present epoch is called the Hubble constant, or Hubble ratio, H_0, and the inverse of this ratio is called Hubble time, t_H. In 1936 Hubble's measurements gave an age of 1.8 billion years for the expansion time scale of the universe—less than the age of the Earth.

Since Hubble's initial measurement, estimates of the expansion rate have shown a predominantly downward trend with a corresponding rise in the estimated age. The first major change in this estimate came in the early 1950s when it was realized that what Hubble had thought to be stars in other galaxies were really whole clusters of stars, a very easy mistake to make. This increased the Hubble time to 8 billion years, an estimate for the first time older than the age of the Earth. More improvements in the distance determinations for nearby galaxies were made in the 1960s and early 1970s, mostly by A. Sandage and G. de Vaucouleurs and their coworkers, almost all increasing the Hubble time until the accepted value for universal age fell between 15 billion and 20 billion years.

Once again, however, measuring galactic distances was not such a simple technique to apply. Gravity is a fly in the ointment. It can slow down the expansion and can even turn it into collapse in the same way that it causes rockets shot spaceward with insufficient velocity to fall back to Earth. In fact, all the mass in the universe acts to slow down the total expansion rate. The effects of this slowdown are shown in figure 5. If there was no matter in the universe, the expansion would continue at a constant rate and the age would equal the Hubble time (case A); but since there is matter in the universe, the age is less than the Hubble time—whether the universe is open and will expand forever (case B) or is closed and will reach a maximum expansion and then collapse on itself (case C). So astronomers need to know both the rate itself and mass-energy content of the universe (to compute the deceleration) in order to estimate its age.

Although there were conflicting measurements of the mass density in the universe on large scales, some of the local measurements (the space around our own galaxy) gave relatively small values for the amount of matter in the universe. This meant that at most only a small correction had to be applied to the Hubble time to get an estimate for the age. Generally as-

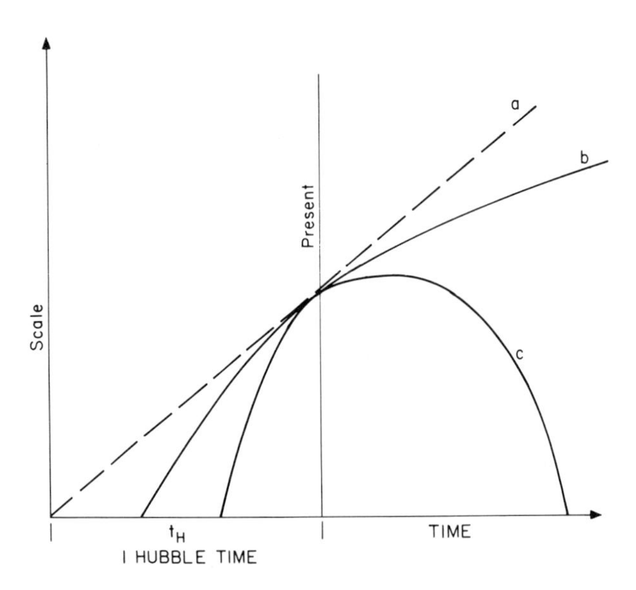

Figure 5
The change of scale of the universe (as measured by the distance between two typical galaxies) with time is given for three possible universes. (a) No matter and no slowdown. The scale increases in direct proportion to time and the universe is open. (b) Some matter and some slowdown. The universe is still open. (c) Enough matter eventually to halt and reverse the expansion; the universe is closed. The Hubble constant, or expansion rate, is just the tangent to the curve at any time. All three cases have the same H_0 and t_H, but in the nonempty cases the true age is less than t_H.

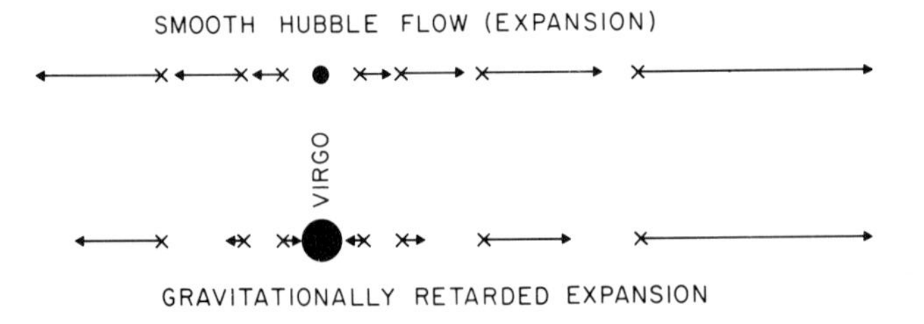

Figure 6
A schematic representation of the expansion of the universe around the local supercluster (Virgo). The bottom diagram shows a very massive Virgo (as we believe it is), greatly slowing the relative expansion of nearby galaxies; the top diagram shows a less massive Virgo, with a small retarding influence. Magnitude of velocity is indicated by length of arrow. In this picture, our galaxy is similar to the third point from Virgo. We are still expanding away from the supercluster center, but our velocity has been slowed two-thirds to three-quarters of its initial value.

tronomers were content with these estimates because they meant that the stellar ages were in rough accord with the universe age.

Unfortunately this complacency has been short-lived. Mass can also slow down the expansion rate locally. Our solar system, for example, is not expanding away from the galactic center. Our galaxy is near the edge of a very large conglomeration of matter centered on the Virgo cluster of galaxies, called the local supercluster of galaxies. Recent measurements of distances to clusters of galaxies by a group of astronomers from the University of Arizona, the Smithsonian Astrophysical Observatory, and Kitt Peak National Observatory have shown that this lump of matter has slowed down the apparent expansion rate rather significantly. An illustration of this effect is shown in figure 6. Because the force of gravity is inversely proportional to the square of distance, points (or galaxies) far away from the central mass are only weakly affected.

The result of this effect from Virgo is that measurements of the Hubble ratio made only locally are incorrect. In order to measure the true expansion rate, it is necessary to measure distances to objects—galaxies or clusters of galaxies—well outside the region affected by the mass of our supercluster of galaxies. Figure 7 shows the best recent calibration of H_0 for

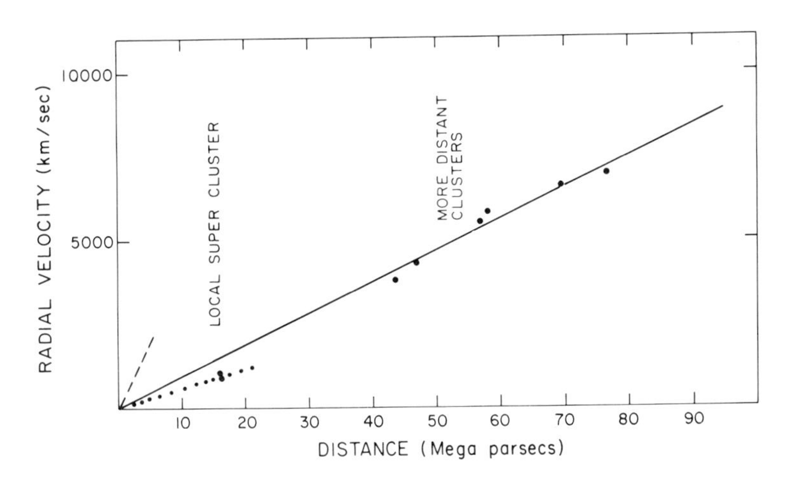

Figure 7
The present calibration of the Hubble constant derived from clusters of galaxies. The short, dashed line near the origin represents Hubble's original calibration. The dotted line below that is the calibration using only two clouds of galaxies inside the local supercluster.

clusters of galaxies outside the local supercluster. This yields a Hubble time of only 11 billion years.

In addition, because it appears that the local supercluster is massive enough to slow the expansion rate substantially, new estimates of the universal mass density are significantly higher than the previously accepted value. They are not sufficient to close the universe but do lower the corrected age estimate to a mere 8 billion years.

Figure 8 summarizes how our estimates of these three time scales, Earth, stellar, and universe age, have evolved over the last one hundred years. The Earth-age measurement has remained steady for the last twenty years, but both the universe and stellar ages have gone up and down. Although it may be small consolation for cosmologists, at least both are still older than the age of the Earth.

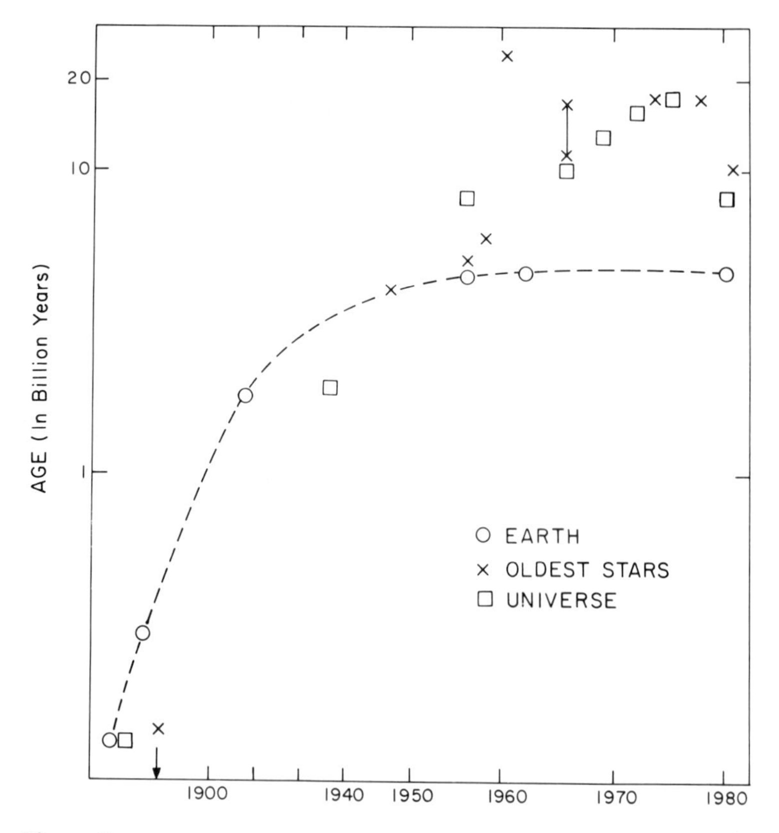

Figure 8
Age estimates for the Earth, stars, and the universe as they have changed over the last one hundred years.

Further Reading

Cloud, Preston. *Cosmos, Earth and Man.* New Haven: Yale University Press, 1978.

Gingerich, Owen, ed. *Cosmology +1.* San Francisco, W. H. Freeman, 1977.

Hubble, Edwin. *The Realm of the Nebulae.* New York: Dover, 1958.

Silk, Joseph. *The Big Bang.* San Francisco: W. H. Freeman, 1980.

Weinberg, Steven. *The First Three Minutes.* New York: Basic Books, 1977.

CHAPTER 8

GEORGE B. FIELD

THREE UNANSWERED QUESTIONS IN ASTRONOMY

How do astronomers really work? As other authors in this book have implied, the advance of scientific knowledge is rarely a neat progression from theoretical prediction to observational proof. In reality, observers rarely wait for questions to be asked; rather they make discoveries by extending their powers of observation with new types of instruments. Thus, x-ray binaries were an answer to a question that, as far as I know, was never asked until after they were discovered. (The question would be, What happens to the matter overflowing the Roche lobe around an evolving O-type star that has a neutron-star companion?)

Nevertheless discoveries quickly lead to new questions. In the case of x-ray binaries, such questions would include, Are the x-rays emitted by hot gas? Why are they pulsed? Is the pulse rate constant, suggesting that they are controlled by the rotation of a compact object? Can the compact object be a white dwarf, or is a neutron star required?

As the simple questions are answered, a conceptual picture or model is built up in the minds of astronomers—in this case, that of an accretion disk of hot gas orbiting a neutron star, with this disk fed by Roche-lobe overflow from the O-type star. The matter of the accretion disk spirals into the neutron star via the poles of its magnetic field, emitting a beam of x-rays, which is observed as pulses as the star spins. Thus further observational work is driven by the desire to sharpen the conceptual picture, and theoretical work contributes by showing how the picture must be modified to explain new observations.

Sometimes this process leads to well-defined questions, which can remain unanswered even after intensive effort. If the question is of broad significance, answering it one way or the other can change the course of further research in a major way. In this chapter, I have posed three questions that seem to be of this type.

Are there more than nine planets in the universe? The first question is posed in this odd way because we have known since 1930 that

Table 1
Discovery of the planets

Planet	Stellar Magnitude	Year of Discovery	Telescope Aperture
Mercury	0	Antiquity	Unaided eye
Venus	−4	Antiquity	Unaided eye
Mars	−2	Antiquity	Unaided eye
Jupiter	−3	Antiquity	Unaided eye
Saturn	+1	Antiquity	Unaided eye
Uranus	+6	1781	15 centimeters
Neptune	+8	1846	20 centimeters[a]
Pluto	+15	1930	33 centimeters

[a]According to Charles Kowal and Stillman Drake, "Galileo's Observation of Neptune," *Nature* (Sept. 25, 1980), Neptune appears to have been observed and noted by Galileo in 1612–1613, when it was only fifteen minutes of arc from Jupiter at the time the great astronomer was observing its satellites. Galileo's telescope was only a few centimeters in aperture.

there are at least nine planets in our own part of the universe, the solar system. Table 1 gives their dates of discovery.

Table 1 teaches us an interesting lesson. We are living on one of the six planets known to the ancients; the other five are visible with the unaided human eye at night. Uranus, of stellar magnitude 6, or about the limit of naked-eye ability, still requires a telescope. In fact, it was discovered accidentally in 1781 by William Herschel from the garden of his home using a 6-inch (15 cm) telescope. (The magnitude scale is a measure of brightness. A difference of one magnitude corresponds to a ratio of brightness of about 2.5, with brighter objects having smaller magnitudes by convention.) By contrast, the discovery of Neptune was the result of a directed search based on theoretical prediction. The mathematical astronomers John Couch Adams and Urbain Jean Joseph Leverrier independently had shown that observed disturbances in the motion of Uranus could not be explained by the gravitational effect of any known planet; rather they required a new body in the solar system whose approximate position could be calculated. On the basis of Leverrier's predictions, the astronomer J. G. Galle discovered the eighth-magnitude planet Neptune with an 8-inch (20 cm) telescope.

The discovery of Pluto was also based on theoretical predictions—except that the theory was wrong. In the early twentieth century, Boston's Percival Lowell, a member of the family that established the Lowell Institute, set up a private observatory at Flagstaff, Arizona, primarily to study Mars. On the basis of his study of the orbit of Uranus, he became convinced that a ninth planet must exist and predicted its general location. His young assistant, Clyde Tombaugh, undertook an extensive search with a 13-inch telescope and, after one year, discovered an object subsequently called Pluto. Strangely our present information on this fifteenth-magnitude planet indicates that it is too small to disturb the orbit of Uranus significantly. In other words, the discovery of Pluto was an accident.

Are there more planets in the solar system? There is no indication from the orbits of the nine planets we know that others are present. Nor have extensive searches with larger telescopes turned up anything. Of course, there could be a very small or very distant tenth planet in the solar system, and we would not know it. A broader question, however, is whether there could be planets around other stars elsewhere in the universe.

Over the years, there has been much speculation on the origin of the planets. Were they captured by the Sun from interstellar space? Were they formed when a passing star pulled a great loop of matter out of the Sun to cool and condense into planets? Or were they the incidental results of a giant, spinning disk of gas and dust, called the solar nebula?

No one knows for sure which, if any, of these concepts is correct, but the last one, based on a solar nebula, seems to explain best many features of our solar system. For example, the fact that all nine planets revolve around the Sun in orbits close to the same plane would be a natural consequence of their accumulation from matter already revolving in a disk. The fact that most of them rotate in the same direction as their revolution is also easily understandable. And the fact that the four inner planets are largely rock while the outer ones appear to be mostly ice, with Jupiter and Saturn having a great deal of hydrogen and helium, seems to be consistent with the idea that the inner nebula would be too hot for ice to form, while the outer nebula would not (figure 1).

Impressed by these and other facts, theorists have developed a detailed theory of the origin of planets, which fits some (but by no means all) of the facts quite well. Fred Whipple and Fred

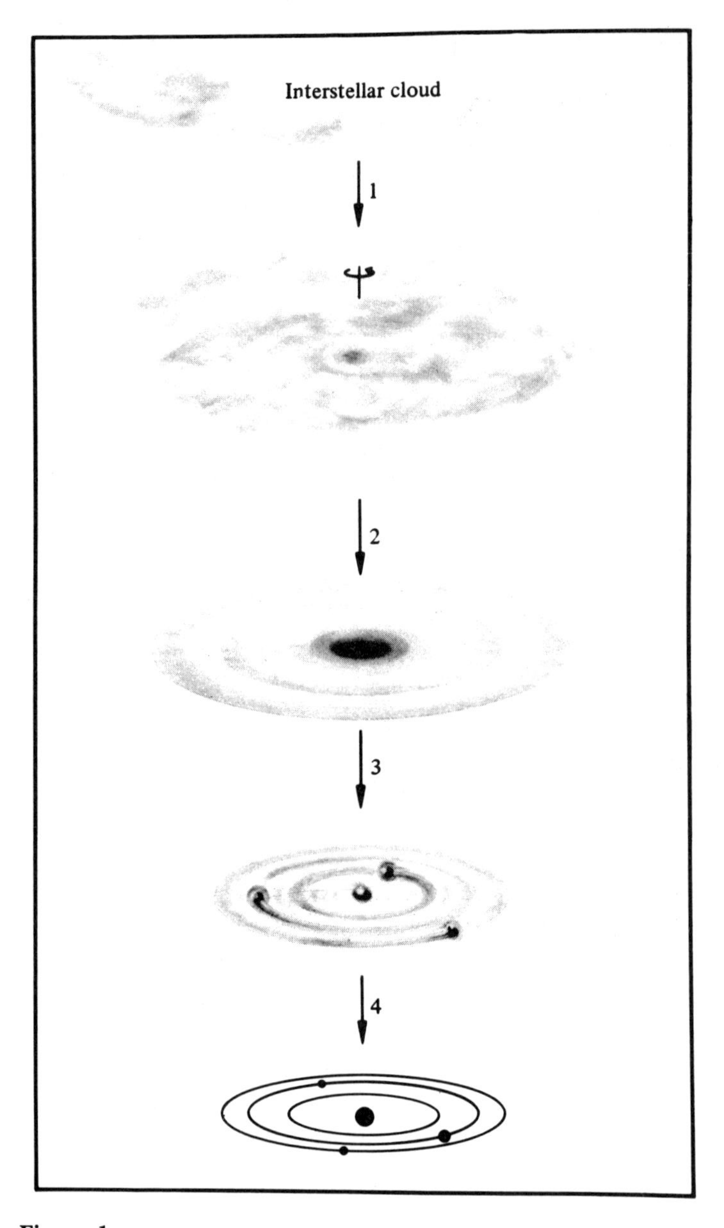

Figure 1
Stages in the formation of the solar system: (1) a fragment of an interstellar cloud collapses; (2) its spin causes it to form a disk; (3) the material in the disk gathers into bodies of planetary size; (4) the Sun, at the center of the system, heats the remaining gas and dust, removing it from the system and leaving the planets in their present form. (From *Cosmic Evolution,* by G. Field, G. Verschuur, and C. Ponnamperuma, Houghton Mifflin, 1978)

Franklin describe this possible theory in this book, but I will review quickly the main points, for they are crucial to our search for planetary systems elsewhere in the universe.

The primordial solar nebula contained perhaps 0.1 solar mass of gas and dust. From the known chemical composition of the Sun and stars, the amount of heavy elements (oxygen, silicon, iron, magnesium) likely to be in the form of dust can be calculated. Then from observations of interstellar dust clouds, the original size of dust grains can be estimated at 10^{-5} centimeters. From the density of grains, a calculation can be made of the time required for them to collide and stick; and from the number of gas molecules (largely hydrogen and helium), the rate at which the larger grains would fall under gravitation to the midplane of the disk can be determined. When the density of the dust grains is found to exceed a certain value, their mutual gravitation would be sufficient to pull them together in kilometer-sized objects. These, in turn, would collide and accumulate to form the inner planets and the cores of the major planets. Jupiter and Saturn probably attracted enough gas from the surrounding nebula to attain their observed sizes and compositions. Unfortunately this theory is not exact enough to predict the distances of the known planets from the Sun or to tell us whether there are unknown planets beyond Pluto. It does, however, suggest trends in the chemical compositions of planets, which can be tested against observation.

What interests us here most, however, is the general characteristic of the theory—its universality. The theory does not require anything special of the Sun or its gaseous surroundings. On the contrary, the general mass, rate of revolution, and chemical composition of the solar nebula are what would be expected when any star like the Sun forms. Therefore the theory predicts that many stars like the Sun should have planets. This prediction, if verified, would open up a whole new branch of astronomy. If we can find other planetary systems, their properties will tell us which properties of the solar system (such as masses, compositions, and distances of planets) are common to various planetary systems and which are accidental in or unique to our system. To progress, the planetary theorist sorely needs more examples of planetary systems.

Astronomers have not yet found a single other planetary system in the universe. Why not? The answer is to be found in the limitations of our present instruments. Even Jupiter, our largest planet, is only a tenth the size and a hundredth the surface area of the Sun. Because of its great distance from the Sun (1000 solar radii) it intercepts only 2.5×10^{-8} of the Sun's radiation and, of that, reflects only about 40 percent. Hence a distant observer on a planetary system elsewhere would see Jupiter as a faint object only a billionth as bright as its parent star. As seen from the distance of even the nearest star (α Centauri, which also happens to be a first-magnitude star very much like the Sun, 4 light-years from Earth) Jupiter would be of twenty-third magnitude; and its orbit would be 4 seconds of arc in radius. Not surprisingly then searches with large telescopes for Jupiter-like objects near other stars have been unsuccessful. Although twenty-third magnitude objects are detectable (though barely) with the largest telescopes, none would be detectable as close as 4 seconds of arc from a first magnitude star such as α Centauri because the light scattered out of the stellar image would completely obscure the planet. All other stars are even farther away than α Centauri, and as a result, any planetary image would be too faint to be detected even with the largest telescopes. (For an example of the obscuration of a faint object by the light of a nearby bright star, see the photograph of Sirius and its white dwarf companion in chapter 2, Kenneth Brecher.)

What, then, can be done? One is to search with the space telescope (ST), a 24-meter diffraction-limited telescope to be launched into Earth orbit on the space shuttle in 1985 by NASA (figure 2). Because it is above the Earth's atmosphere, ST will produce much sharper images, perhaps only 0.″025 in radius. This should help in the search at least for large planets with large orbits because the light from both the star and the planet then would be more concentrated and more easily resolved into two distinct images.

But more also can be done from the ground. As Jupiter swings around the Sun on its twelve-year orbit, its gravitational pull displaces the Sun by about 1 solar radius. From a distance of 30 light-years, astronomers would see this as a periodic displacement of the stellar image by 0.″0005. Even though atmospheric blurring smears the stellar image over a much larger area, because of recent improvements in technology, the position of its center can be determined with a precision of ±0.″001, or only somewhat larger than the value required for detection of the perturbations caused by a Jupiter-like planet around a

Figure 2
An artist's conception of the space telescope, scheduled for launch in 1985. (Courtesy NASA)

star 30 light-years distant. Furthermore, for relatively modest cost, the precision could be increased to $\pm 0\overset{''}{.}0001$ (figure 3).

Why are we so interested in planetary systems in the first place? After all, less than 1 percent of the material in the solar system ended up in these cold chunks of matter orbiting the Sun. The answer, perhaps, is life. Throughout the ages humankind has speculated about life on other worlds. Unfortunately the other worlds in our own planetary system seem barren. For example, Mars must have had water once, but any remaining moisture seems locked in permafrost, and there is no sign of life at the random spots where the Viking landers put down (figure 4). And the best guess is that no other planet in the solar system has any better chance than Mars of harboring life.

Beyond the solar system, the question of life is completely open. Certainly the conditions on a rocky planet orbiting another Sun-like star could well support life with light, liquid water, carbon, nitrogen, and trace elements. But given the difficulty of detecting planets, the chance of making direct observations of life forms seems hopeless. If the life forms were intelligent, it would be different, for intelligent life, at least on Earth, has a way of making its presence known over interstellar

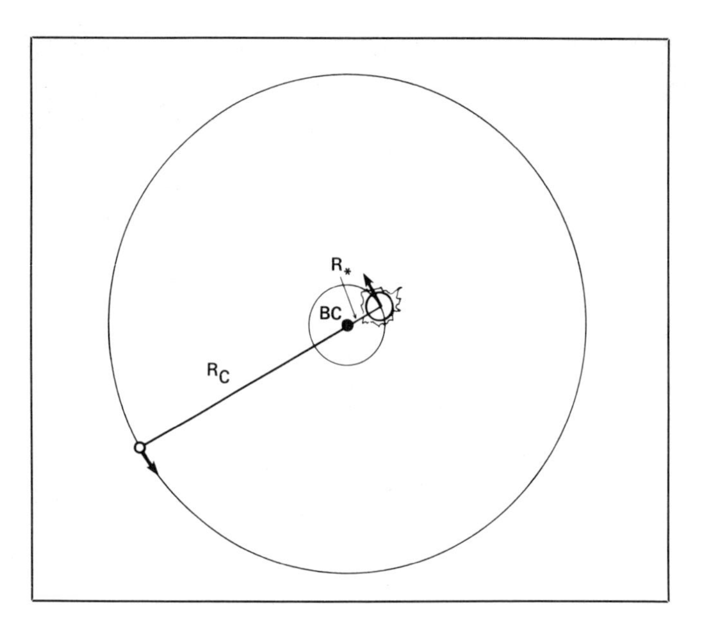

Figure 3
A schematic representation of a simple planetary system. Both the planet (small open circle) and the star (large open circle) orbit their common center of mass. Even though the planet may be invisible in the telescope, the resulting motion of the parent star can be detected. (Diagram from *Mercury*; copyright © 1980, Astronomical Society of the Pacific)

distances. Since the 1930s, human-generated radio signals—mostly in the form of commercial radio and television broadcasts, including soap operas, game shows, and sports events—have been propagating into interstellar space at the speed of light. Today the first radio signals are about 50 light-years distant, and they have probably reached some ten thousand stars. In addition to these accidental messages, a coded communiqué sent in 1974 from the world's largest radio telescope at Arecibo, Puerto Rico, and powerful enough to be detected by another Arecibo on the other side of the Milky Way, has now traveled 6 light-years into space (figure 5).

Assuming that similar signals from other worlds might be directed our way, Frank Drake began a search for extraterrestrial intelligence (SETI) in 1961. So far these attempts have been without success. But the efforts have been at a very low level, constrained by the ability of radio astronomers to bootleg time on big telescopes. It has been proposed to start a serious

Figure 4
The surface of Mars as seen from Viking I. There is no sign of life at this site. (Photograph from NASA)

effort: a full-time search with a specially designed receiver a million times more sensitive than any used so far. Who knows what SETI might find? Even if only primitive signals are identified, we would at least identify the parent star, the orbit of the inhabited planet, and perhaps some of its characteristics. This could be another way to find out about planets. The discovery that we are not alone in the universe would totally eclipse such narrow scientific considerations, of course. As is often the case in astronomy, the discovery would raise far more questions than it answers.

Is the theory of stellar evolution wrong? No one has ever seen below the Sun's surface. But just as seismologists have learned about the interior of the Earth from earthquakes, astronomers have learned about the interior of the Sun from observations of its surface. To make sense of these observations, astronomers use theoretical models of the Sun and stars based upon simple physical principles. The first is that the inward force of gravity must be balanced by the outward push of pressure associated with the high temperature of the solar interior. The second is that the same high temperature drives energy through the outer layers of the Sun, to emerge at its surface as sunlight, and the

Figure 5
The 300-meter radio telescope constructed in a natural depression in the hills near Arecibo, Puerto Rico. (Courtesy National Astronomy and Ionosphere Center, operated by Cornell University under contract with NSF)

rate of this energy transfer can be calculated from the properties of hot gases. The third is that the energy thus lost from the interior must be made up by the nuclear energy released when four hydrogen nuclei (protons) fuse to form a single helium nucleus; this too is a calculable process.

These principles, expressed in precise mathematical form, lead to equations that can be solved to yield predictions for the Sun's radius and luminosity, both of which can be observed. Although there are uncertainties in the theoretical assumptions (such as the rate at which hot gases absorb radiation), the most reasonable assumptions lead to agreement with observations of the solar luminosity and radius. Does this mean the actual solar interior is like that of our models, with a central density about one hundred times that of water, and a central temperature of about 15 million degrees? Perhaps. But there is a major problem.

For a given solar model, the number of electron neutrinos (elusive subatomic particles) that must be produced as a by-product of the nuclear reactions in the solar interior can be calculated. Since neutrinos are so penetrating that they can easily escape directly from the interior of the Sun, they should be detectable at Earth; however, an experiment to detect them disagrees substantially with prediction, and this has led to what has been called the solar neutrino problem. Indeed the discrepancy between theory and observation has far-reaching implications because the same models for describing the Sun have been applied to all stars and are the basis for our ideas about stellar evolution.

The ability of neutrinos to penetrate matter makes them extremely difficult to detect. In fact, to do so, Ray Davis of Brookhaven National Laboratory has had to exert great efforts over ten years. His experiment consists of placing a large tank containing 100,000 gallons of tetrachloroethylene, a common cleaning fluid, some 1500 meters deep in the Homestake Gold Mine, at Lead, South Dakota (figure 6). (The great depth is to shield the experiment from all particles but solar neutrinos.) The active element of the experiment is chlorine and about 25 percent of natural chlorine is the isotope ^{37}Cl, whose nucleus contains seventeen protons and twenty neutrons. A solar neutrino of sufficient energy has a finite probability of striking one of the neutrons in ^{37}Cl and converting it to a proton; the result is an argon nucleus of eighteen protons and nineteen neutrons (^{37}Ar).

Figure 6
The solar neutrino experiment, Homestake Gold Mine, Lead, South Dakota. The tank contains tetrachloroethylene, a common cleaning fluid. (Photograph from Brookhaven National Laboratory)

The nucleus of ^{37}Ar is unstable; in a period of typically thirty-four days, one of its protons reaches out to interact with one of the electrons orbiting the nucleus and combines to form a neutron again and, in the process, emits energy. Thus Davis periodically flushes out his tank looking for any argon produced by solar neutrinos. Over the years he has perfected the technique so that he can locate single atoms. A solar neutrino flux production rate of one argon atom every forty days in his tank is defined as a solar neutrino unit or SNU. (Since scientists are no less susceptible to bad punning than anyone else, the query, "What SNU?" is inevitably heard at scientific conferences.)

The best theoretical prediction for the solar neutrino flux is

7 or 8 SNU; the best estimate of the experimental counting rate in Davis's tank is 2 or 3 SNU, about a factor of three smaller. Is this a disaster for astrophysical theory, or can the discrepancy be easily explained away? We do not know.

Some of the parameters assumed in the theory are uncertain. Furthermore, the side branch in the nuclear reaction chain in the Sun that produces neutrinos energetic enough to be captured by chlorine is highly temperature sensitive. Thus with a few changes in the assumptions regarding absorption coefficients, nuclear reaction rates, or the efficiency of thermal convection in the Sun's outer layers, we could change the temperature calculated for the solar interior enough to bring theory into line with experiment. On the other hand, the theory as it now stands has been used to describe the interiors of all types of stars. More important, this theory of stellar interiors predicts the shape of the Hertzsprung-Russell diagram (see figures 2–4 in chapter 7, John Huchra) from which the ages of stellar clusters have been derived. In short, virtually all of what we assume about stellar evolution has been called into question by the solar neutrino problem, and solving it is crucial for astronomy.

For these reasons it is desirable to conduct other neutrino experiments. Since the element gallium (in particular, the isotope ^{71}Ga) is sensitive to neutrinos of lower energy, produced in a nuclear reaction that is not so temperature dependent, a gallium detector could directly test the hypothesis that solar energy is produced by fusion of hydrogen to helium, independent of the temperature of the solar interior. A gallium detector is being built now at Brookhaven; it is a collaborative project involving groups in Germany and Israel and at Princeton's Institute for Advanced Study and the University of Pennsylvania.

If our theory is basically correct, except for one or more of the assumed parameters, the gallium experiment will tell us and identify the new range of parameters. Standard theory assumes that the Sun formed from an interstellar cloud whose chemical composition was uniform. However, it is possible that at various times over its 4.5-billion-year life, the Sun has passed through dense interstellar clouds and thereby accreted matter on its surface that is somewhat richer in heavy elements than is its interior. If so, the interior of the Sun may have a somewhat lower abundance of heavy elements than does the outermost layer we observe. This would result in a cooler interior, because the solar gases would then be more transparent than theorists have assumed.

Standard theory also assumes that the electron neutrino is a simple particle. This assumption stems from nuclear physics, where until recently it has been the simplest explanation for a large number of phenomena. The suspicion has grown, however, that the electron neutrino bears an extremely intimate relationship to its cousins, the μ neutrino and the τ neutrino—so intimate, in fact, that each type changes into the other and back again within a millionth of a second. Such neutrino oscillations, which are suggested by some contemporary theories of particle physics, may be supported by experiments with neutrinos at nuclear reactors. These experiments, however, are very difficult and have not yet been confirmed. If they should turn out to be correct, they would have immediate consequences for the solar neutrino problem. For example, even before leaving the Sun, the electron neutrinos could be converted to a random mixture of electron, μ, and τ neutrinos. Since the chlorine in the Davis experiment is sensitive to only one of the three types, the factor-of-three discrepancy would be immediately explained. In fact, the gallium experiment could take on added significance because its capture rate can be accurately predicted from only the observed total energy production rate in the Sun, without any detailed knowledge of the interior temperature. Any serious discrepancy between the gallium experiment and the theory would then have to be attributed to neutrino oscillations, which themselves are of great interest to physicists.

In fact, neutrino oscillations could have other rather dramatic implications for astronomy. It is believed that neutrinos, like the photons we now see as the cosmic microwave background radiation, were created in large numbers during the first three minutes in the life of the universe. Normally it is assumed that neutrinos, also like photons, have a zero rest mass and therefore must always move at the speed of light. But if neutrino oscillations are found, the neutrino rest mass must be finite. Present laboratory results imply only that the rest mass is very much less than that of an electron. But since galaxies could be affected by the gravitational attraction of such heavy neutrinos, then from the observed motions of the galaxies, we can deduce that the neutrino rest mass must be less than 3×10^{-4} that of the electron. Even so small a value would be of great interest to astronomers because the heavy neutrinos would make up much of the mass in the universe now thought to be invisible.

In what form is the hidden mass in galaxies? Since Newton it has been possible to derive the mass of an astronomical body (that is, the total amount of matter within it) from observing the motion of a second body in its gravitational field. Thus the mass of the Sun can be derived from the orbit of any of its planets, and the mass of any planet can be derived from the orbit of any of its satellites.

Briefly the method consists of equating the centrifugal acceleration acting on a body in a circular orbit (proportional to the square of the orbital velocity and inversely proportional to the radius of the orbit) to the gravitational acceleration on the body (proportional to the mass of the gravitating body and inversely proportional to the square of the radius of the orbit). Thus the mass of the gravitating body must be proportional to the square of the orbital velocity and to the radius of the orbit. Both of these quantities can be easily observed in the solar system since the periods and radii of planetary and satellite orbits are known from direct observation, and the orbital velocity is simply the circumference of the orbit divided by the period. From just such calculations, we know the mass of the Sun is 2×10^{33} grams.

Remarkably enough, astronomers also have been able to weigh many other stars in spite of their enormous distances from Earth. The method depends on discovering pairs of stars that are in orbit around each other. In some cases, the radius of the orbit is large enough to resolve with a telescope so that the astronomer sees two individual stars. By watching them over a long time, the observer can determine an orbital period, and if the distance of the pair can be found (so that the angular separation of the two stars can be converted to an orbital radius), the masses can be determined.

More often, however, the two stars are so close that the two images merge into one. Yet, surprisingly, even in such cases, it is possible to deduce stellar masses. When the combined light from the pair is passed through a spectrograph, the spectra of the individual stars can be separated, and a Doppler shift arising from the orbital motion can be observed. (See chapter 5, Harvey Tananbaum, for an explanation of the Doppler shift.) Thus the orbital velocity is directly observed. Because the shifts oscillate

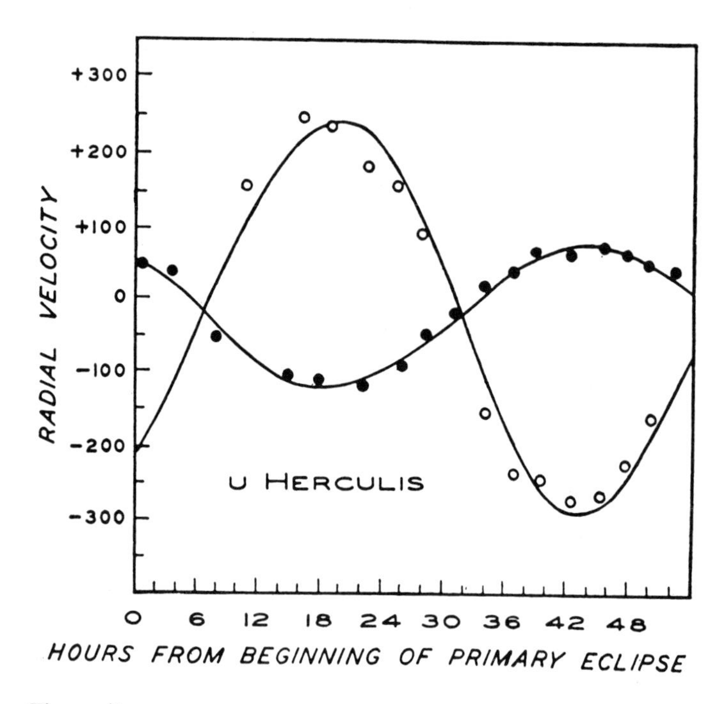

Figure 7
Radial velocities determined in the eclipsing binary system μ Herculis. From the amplitude of the velocities and the period of the system, the masses can be determined.

back and forth as each star first approaches and then recedes from us in its orbital motion, we can obtain the orbital period (figure 7). Even though the orbital radius cannot be directly observed in such cases (since by definition the stellar images cannot be resolved), it can be calculated from observations of the orbit velocity and of the period. Then the mass can be deduced by using velocity and radius.

With this technique, it has been shown that stars of nearly the same luminosity and temperature as the Sun have nearly the same mass, as predicted by the theory of stellar structure. It has also been found that many stars form a natural sequence, with the stars more massive than the Sun being brighter and hotter and those less massive being fainter and cooler. The existence of this main sequence is also predicted theoretically from calculations of the rate at which stars burn their hydrogen into helium (see figure 2, chapter 7, John Huchra).

But not all stars fall on the main sequence. For stars of a given temperature, there are also stars that are much brighter ("giants") and much fainter ("white dwarfs") than the Sun. They, too, are natural stages in stellar evolution. More recently, stars too faint to be seen have been detected by completely different means. Neutron stars are observed either as radio or as x-ray sources. Their masses are, as expected, comparable to that of the Sun. And in one case, the remarkable x-ray source Cygnus X-1, it can be inferred that a black hole must be present that emits no light at all. This mass is calculated to be about 10 solar masses, as determined from the motion of its visible companion.

Thus five types of stars are known: giants, main-sequence stars, white dwarfs, neutron stars, and black holes. Although their visual luminosities depend on their state of evolution and surface temperature, typical luminosities in solar units are of the order of 10,000, 1, 1/10,000, negligible, and zero, respectively. Thus, in any mixture of stellar types, the light tends to be dominated by giants, with some contribution by main-sequence stars, and only negligible contributions by white dwarfs, neutron stars, and black holes.

These ideas can be put to the test by observing globular clusters. Here hundreds of thousands of stars evidently orbit a common center under the influence of their mutual gravitation. (See figure 1, chapter 7, John Huchra.) The mass of the cluster can be inferred from its observed size, and the orbital velocities of its members can be determined by Doppler shifting of the spectra. The mass can be compared with the observed luminosity by calculating a mass-luminosity ratio (M/L), defined as the mass (in solar units) divided by the luminosity (also in solar units). The typical result for a globular cluster ($M/L \simeq 0.8$) suggests that the cluster contains primarily stars like the Sun. But this cannot be true because the integrated spectrum of globular clusters is similar to that of a giant star, not the Sun, thereby suggesting that giants contribute most of the light. On the other hand, M/L for giants is only about 10^{-4}. This enigma is resolved by painstaking observations of each of several hundred stars in a cluster. When these stars are plotted on a temperature-luminosity diagram (as in figure 3, chapter 7, by Huchra), a well-defined sequence of stars is found, with both giants and main-sequence stars present. Although only the brighter main-sequence stars are usually detected, the numbers increase rapidly as one goes to fainter stars, much as has been found to be the case with the stars near the Sun. When the relative numbers of giants and main-sequence stars are compared, the observed M/L ratio can be explained: the cluster must be a composite system in which most of the mass is contributed by large numbers of relatively faint main-sequence stars but in which most of the luminosity is contributed by a few relatively bright giants. Of course, this also explains why the composite spectrum is like that of a giant.

This success in describing the composition of globular clusters suggests applying the same approach to galaxies. Like globular clusters, galaxies appear, at least superficially, to be made up of large numbers of stars. Their spectra can be explained as a composite of various types of stars, largely giants. And in the nearest galaxies, such as the Magellanic Clouds and the Andromeda Nebula (Messier 31), the brightest stars can be individually resolved, and they do appear to be like the giants in our own galaxy.

It is helpful to make a distinction between spiral galaxies like our own or Andromeda, which appear to be huge disks of stars, interstellar gas, and dust in orbit around a relatively gas- and dust-free nucleus (figure 8) and the much rarer, giant, spherical galaxies like Messier 87, which appear to be relatively free of gas and dust and composed of stars distributed in a spherical volume (figure 9).

At first the spherical galaxies seem to resemble globular clusters. Because they are too distant for us to resolve the individual stars, information gathered along different lines of sight at various projected distances from the galactic center is utilized. Again the Doppler spreading of spectral features yields the orbital velocities of stars, which can be combined with the radial distance to give the mass contained within each radius or, equivalently, the mass density at various radii. The variations in luminosity density are also determined from the way the brightness decreases with projected distance.

When the ratio of mass density to luminosity density is calculated close to the galactic center, M/L is of the order of 10 solar units. This value is quite close to the composite for stars near the Sun, suggesting that the stellar population may be similar to that near the Sun. However, the situation at greater distances from the center is anomalous. Outside the core of the galaxy, the luminosity density decreases like the inverse cube of the radius, while the mass density appears to decrease like the

Figure 8
The Andromeda Nebula, Messier 31. (Courtesy Palomar Observatory, California Institute of Technology)

Figure 9
The giant spherical galaxy Messier 87 in the Virgo cluster of galaxies. Note the globular clusters arranged around it. (Courtesy Kitt Peak National Observatory)

inverse square of the radius. It follows that the M/L ratio increases in proportion to the radius. This fact alone proves that spherical galaxies cannot be a uniform mixture of stars of various types.

One must also study how the spectrum changes with distance from the center. When this is done, two results emerge. First, the spectrum is similar to giant stars at all distances, and second, other subtle variations with distance indicate that the abundance of heavy elements decreases with increasing radius. The simplest interpretation is that there is a steady change in stellar population with radius. Since the mass is inferred from Newton's laws but is not directly observed, the variation of M/L in spherical galaxies is an example of what has become known as the hidden mass problem, one of the key questions in astronomy today.

Earlier I mentioned spiral galaxies. In such cases, the disk of the galaxy is observed to be rotating about the center of the galaxy, much as planets orbit the Sun. However, the observed orbital velocities do not decrease with radius, as would be expected if all the mass were concentrated at the center, as in the solar system. Rather the velocity tends to remain constant with radius out to the largest distance at which it can be observed. Only recently, with the perfection of radio astronomy techniques and new spectrographs on large optical telescopes, has it been possible to observe this phenomenon at distances greatly exceeding that of the Sun from the center of the Milky Way galaxy, typically about 10,000 parsecs. (A parsec is about 3 light-years.) In many galaxies the phenomenon has been observed out to 20,000 parsecs and, in a few, out to 40,000 parsecs or more. In no spiral galaxy observed does the orbital velocity decrease significantly at large radii. This result requires that the mass within a given radius increase in proportion to the radius.

Where is the matter responsible for the observed gravitational effect? Conceivably it could be located in the disk of the galaxy, but it has been shown mathematically that any disk this massive would be violently unstable. For this reason, it has been suggested that the mass responsible is actually distributed, as in the case of a spherical galaxy like Messier 87, in a spherical volume centered on the center of the galaxy. In such a case, one can calculate how the mass density should decrease with radius in order to account for the observations. Interestingly enough, it must decrease inversely as the square of the radius, just as was found in spherical galaxies. Thus spiral galaxies are suspected to be similar to spherical galaxies in some respects. In both, most of the mass is distributed in a spherical volume. In spherical galaxies that is all there is, but in spiral galaxies there is also a disk of material that contains only a small fraction of the mass but (because of its luminous stars) is much more readily visible than the spherical distribution.

There are other ways to determine the masses of galaxies. Like stars, galaxies are sometimes observed in pairs, which are presumably orbiting each other with enormous periods (10^9 years or more). The Doppler shifts corresponding to the expected orbital velocities are readily observable; by combining them with the observed orbital radii, masses can be calculated. The corresponding M/L ratios are of the order of fifty, quite consistent with the increase of M/L with radius observed for isolated spherical and spiral galaxies.

Galaxies often occur in groups of a dozen or so, which, in turn, are parts of giant clusters of galaxies having a thousand or more members (figure 10). Again the total mass and lumi-

Figure 10
The Coma cluster of galaxies, a system containing over 1000 members, of which the two supergiant ellipticals are most evident in this picture. In addition, Coma contains large amounts of intergalactic gas at a temperature of 100 million degrees, invisible in this picture but seen with x-ray telescopes. (Courtesy Kitt Peak National Observatory)

nosity of the system can be determined. Values range from about 50 up to about 300, continuing the trend to the enormous distances (100,000 parsecs and more) between the galaxies in such systems.

Finally, there are approximate methods of determining the amount of mass distributed over even greater distances—10^6 parsecs and more. These methods involve applying the usual relation between radius and velocity to a statistical collection of distances and velocities of galaxies to obtain their masses. The result is that M/L averages about 250 for large volumes of matter, which appears to be consistent with results from the other methods and has fascinating cosmological implications. From the observed numbers of galaxies and their typical luminosities, one can calculate that each cube 10^6 parsecs on a side (a cubic megaparsec) typically contains a luminosity of 4×10^8 times that of the Sun. If M/L is 250, the same cube contains 10^{11} solar masses.

According to the General Theory of Relativity, the expansion of the universe is constantly decelerated by the gravitational effect of the matter it contains. In fact, for the best current estimate of the Hubble constant of expansion (see chapter 7, John Huchra)—90 kilometers per second per megaparsec of distance—the density of matter required to halt and reverse the expansion at some time in the future is 2×10^{11} solar masses in a cubic megaparsec. Thus current evidence from the dynamics of galaxies suggests that the amount of matter associated with galaxies is roughly half that required to stop the expansion. Is it just a coincidence that this number is so near unity, when it could equally well have been one hundred or one one-hundredth? Could it actually be unity, and we have made errors in finding its value to be one-half? No one yet knows the answers to these questions.

A more pertinent question concerns whether the hidden mass takes the same form in all galaxies. In view of the similarity of its distribution in various types of galaxies, we may make the simple hypothesis that it does. But what form?

Possible forms of hidden matter are suggested by the standard theory of stellar evolution. Stars are born out of interstellar gas, burn their hydrogen as main-sequence stars, and then, if they are more massive than the Sun, evolve through a giant phase into white dwarfs, neutron stars, or black holes, depending on their initial mass. Giants are too bright to account for hidden mass, but all the other steps in this evolutionary pattern should be considered.

Interstellar gas. Searches for hydrogen atoms and molecules in the outer parts of galaxies have failed, so any hydrogen present must be ionized (broken into protons and electrons). Recently, orbiting ultraviolet telescopes have discovered such gas in our galaxy and in the Magellanic Clouds, but the amounts of gas involved are negligible.

Main-sequence stars. M/L for individual stars increases rapidly as one goes to fainter stars, reaching 100 only for stars whose surface temperature is less than 2400 degrees Kelvin. Such stars, of spectral type M8, are very numerous near the Sun, but they are not numerous enough to make the overall M/L approach 100; in fact, the local value averaged over all stars is only about 8. To explain the M/L in the outer parts of galaxies, there must be a rapid increase in the relative numbers of faint main-sequence stars as one goes away from the center of the galaxy. This would happen if for some reason the stars that form in the outer parts of the protogalaxy have smaller masses on average than those in the inner parts; but no one knows why this should be so.

White dwarfs, neutron stars, black holes. These candidates are similar in having negligible luminosities and in being the end products of stellar evolution. It is known that white dwarfs, which would be expected to contribute more mass than do neutron stars or black holes, contribute less than one-sixth of the total mass near the Sun. To generate $M/L = 100$ would require a seventy-fold increase in the number of white dwarfs. Since white dwarfs are formed as the result of the evolution of massive stars, the outer parts of galaxies would have to have far more massive stars than are found in the solar neighborhood. Such massive stars would eject the products of nucleosynthesis into space, so that more heavy elements in the outer parts of galaxies would be expected, rather than less, as is observed. Thus white dwarfs appear to be an unlikely explanation for hidden mass, and neutron stars and black holes are even less likely ones. Of the four types of stars other than giants, only faint main-sequence stars seem candidates for contributing the hidden mass.

There is, however, another possibility: *massive neutrinos.* The solar neutrino problem might be solved if we could find evidence for neutrino oscillations. Recall that neutrino oscillations can occur theoretically only if the neutrino rest mass is more than zero. From the physical conditions existing in the early universe, we suspect that the number of neutrinos must be about equal to the number of photons in the cosmic background radiation. Since the latter is directly observable, one can calculate that there must be about 100 neutrinos per cubic centimeter throughout the universe, or 3×10^{75} neutrinos per cubic megaparsec. The hidden mass in such a volume is 10^{11} solar masses, or 2×10^{44} grams. Suppose that the hidden mass were made up of massive neutrinos. Then the rest mass of each neutrino would have to be 2×10^{44} grams divided by 3×10^{75}, or 7×10^{-32} gram, which is equal to 1.3×10^{-4} electron mass. Such an extremely small mass cannot now be measured in the laboratory, so there are no experiments either to confirm or to deny such a value.

If conventional physics is correct, the most likely explanation for the hidden mass of the universe is faint main-sequence stars, whereas a possible hypothesis based on new physics would be that it is in massive neutrinos. How can astronomers find out which hypothesis is correct? This seems like a very difficult task, since the mass is by definition hidden. We must think of new ways to detect these forms of matter.

One approach is based on the fact that examples of faint main-sequence stars near our solar system are active; that is, they have emission lines like those of the solar chromosphere, x-ray emission like that of the solar corona, and flares like those on the solar surface. These phenomena are thought to be manifestations of magnetic fields generated by turbulent motions within the star and whose energy dissipates to heat the outermost layers of the star. In short, a significant fraction of the total energy produced by the star is released not in the form of the infrared radiation, which is characteristic of the low surface temperature of the star, but in the ultraviolet and x-ray parts of the spectrum characteristic of gases in the 100,000 to 1 million degree range. This fortuitous fact might make large numbers of such stars located in the outer parts of galaxies visible as a diffuse glow of ultraviolet or x-rays, particularly inasmuch as the background in those parts of the spectrum against which the radiation must be detected is much lower than in the infrared.

It is much harder to test observationally the neutrino hypothesis; however, it has been suggested that massive neutrinos would be subject to decay. The most likely decay mode would be into two photons, each of which would carry away an energy equal to half the rest mass. The peak of the expected light distribution would be at about 400 angstrom units, in a part of the spectrum that can be observed only by spacecraft. (Because any radiation at wavelengths shorter than 912 angstrom units would be attenuated by interstellar hydrogen, holes must be found in the interstellar medium through which to observe.) Observations of this type are being planned.

There is also an approach based on the theory of the early universe. According to the big-bang model, almost all the free neutrons produced in the big bang combined with protons to form deuterons, the nuclei of the chemical element deuterium. In turn, most of the deuterons reacted further to form helium, a result we see today in that 25 percent of the matter in stars and interstellar space is helium. However, some of the deuterons escaped this reaction in the rapidly expanding early universe and should now be present as deuterium atoms. In fact, these atoms are observed today in interstellar space. Their number is about 1.4×10^{-5} of the number of hydrogen atoms. According to the theory, the deuterium-hydrogen ratio is controlled by the density of normal matter in the universe because the greater the density of matter, the less deuterium would escape reacting. Since some deuterium may be destroyed in stars, the observed interstellar abundance yields a lower limit on the abundance of deuterium produced in the big bang, and hence an upper limit on the amount of normal matter in the universe. This upper limit is only about 20 percent of the amount of hidden mass inferred by observations of galaxies. Thus there is evidence that whatever the hidden mass is, it cannot be normal matter. This argument appears to rule out faint main-sequence stars while allowing massive neutrinos to comprise the hidden mass.

By no means has the sky yielded up all its mysteries. Even if we put aside the profound question of cosmic intelligence, there remain questions whose answers would surely unleash new efforts in the pursuit of knowledge. When the tenth planet is reported, our chances of attaining a correct theory of the origin of the solar system will increase. When the gallium experiment yields the flux of low-energy solar neutrinos, we will know whether it is the basic physics or the theory of stellar evolution that needs to be revised. And when new techniques are applied to studies of the outer parts of galaxies, we will know what form most of the matter in the universe takes and therefore be challenged to develop new theories of the universe. As we have witnessed throughout history, the dynamic interplay of prediction and proof will press forward the frontiers of science.

Further Reading

Bahcall, J. N. "Neutrinos from the Sun." *Scientific American* (July 1969).

Faber, S. M., and Gallagher, J. S. "Masses and Mass-to-Light Ratios of Galaxies." *Annual Reviews of Astronomy and Astrophysics* 17 (1979): 135.

Field, G. B. "The Mass of the Universe: Intergalactic Matter." In *Frontiers of Astrophysics,* edited by E. H. Avrett. Cambridge: Harvard University Press, 1976.

Field, G., Verschuur, G., and Ponnamperuma, C. *Cosmic Evolution.* Boston: Houghton Mifflin, 1978.

Fowler, W. A. "The Case of the Missing Neutrinos." *Science Year 1974,* pp. 78–91.

Iben, I., Jr. "Globular-Cluster Stars." *Scientific American* (July 1970).

Morrison, P., Billingham, J., and Wolfe, J., eds. *SETI: The Search for Extraterrestrial Intelligence.* Washington: NASA, 1977.

Sandage, A. R. "Cosmology: A Search for Two Numbers." *Physics Today* (February 1970), pp. 34–41.

CONTRIBUTORS

Kenneth Brecher is professor of astronomy and physics at Boston University. Brecher did both his undergraduate and graduate work at the Massachusetts Institute of Technology, receiving the Ph.D. in 1969. His current research interests include high-energy astrophysics, relativity, cosmology, and archaeoastrophysics. Brecher is coauthor of *High Energy Astrophysics and Its Relation to Elementary Particle Physics* (MIT Press, 1979) and coeditor of *Astronomy of the Ancients* (MIT Press, 1979).

George B. Field is director of the Harvard-Smithsonian Center for Astrophysics, a cooperative research venture combining the facilities and staffs of the Harvard College and Smithsonian Astrophysical observatories. Field is also Paine Professor of Practical Astronomy at Harvard University, and his research interests include the study of interstellar and intergalactic matter and cosmology. Field did his undergraduate work at the Massachusetts Institute of Technology and received the Ph.D. from Princeton University in 1955. He is the coauthor of *The Red Shift Controversy* (W. A. Benjamin, 1974) and *Cosmic Evolution* (Houghton Mifflin, 1978), and coeditor of *The Dusty Universe* (Neale Watson Academic Publications, 1975).

Fred Franklin is an astronomer on the staff of the Smithsonian Astrophysical Observatory and a research associate of the Harvard College Observatory. Franklin took both his graduate and undergraduate degrees at Harvard, receiving the Ph.D. in 1962. His research interests include the study of asteroids, planetary ring systems, and comparative planetology.

Owen Gingerich is an astrophyscist at the Smithsonian Astrophysical Observatory and a professor of astronomy and the history of science at Harvard University. An authority on Copernicus and Renaissance astronomy, Gingerich is also interested in studies of solar and stellar atmospheres. He did his undergraduate work at Goshen College and graduate work at Harvard, receiving the Ph.D. in 1962. He is the editor of *New Frontiers in Astronomy* (W. H. Freeman, 1975) and *Cosmology +1* (W. H. Freeman, 1977) and coeditor of *A Source Book in Astronomy and Astrophysics* (Harvard University Press, 1979).

John Huchra is an astronomer on the staff of the Smithsonian Astrophysical Observatory and a research associate of the Harvard College Observatory. Huchra did undergraduate work at the Massachusetts Institute of Techology and received the Ph.D. from the California

Institute of Technology in 1976. An observational astronomer who divides his time between Cambridge and observatories in the American Southwest, Huchra is interested in cosmology and extragalactic astronomy.

Walter H. G. Lewin is professor of physics at the Massachusetts Institute of Technology. Born and educated in the Netherlands, he received the Ph.D. from the University of Delft in 1965. His research specialty is high-energy astrophysics, particularly x-ray observations conducted from satellites and balloons. In 1978, Lewin received a NASA award for Exceptional Scientific Achievement for his satellite experiments in x-ray astronomy.

Alan P. Lightman is a physicist at the Smithsonian Astrophysical Observatory and a Lecturer in astronomy and physics at Harvard University. Lightman did his undergraduate work at Princeton and his graduate studies at the California Institute of Technology, receiving the Ph.D. in 1974. Lightman is coauthor of two books: *Problem Book in Relativity and Gravitation* (Princeton University Press, 1975) and *Radiative Processes in Astrophysics* (John Wiley, 1979). His research interests include the study of black holes, quasars, radiative processes, and stellar dynamics.

Robert W. Noyes is an astrophysicist at the Smithsonian Astrophysical Observatory and professor of astronomy at Harvard University. From 1973 to 1980, he also served as associate director for solar and stellar physics at the Center for Astrophysics. Noyes did his undergraduate work at Haverford College and received the Ph.D. from the California Institute of Technology in 1963. The author of *The Sun: Our Star* (Harvard University Press, 1981), Noyes is interested in studies of solar physics and ultraviolet solar and stellar astronomy.

William H. Press is professor of astronomy and physics at Harvard University. His research interests range from gravitational waves and black holes to general studies in cosmology. Press is the coauthor of *Problem Book in Relativity and Gravitation* (Princeton University Press, 1975). He did his undergraduate work at Harvard University and his graduate work at the California Institute of Technology, receiving the Ph.D. in 1972.

Robert Rosner is assistant professor of astronomy at Harvard University. He received the B.A. from Brandeis University in 1969 and the Ph.D. from Harvard University in 1975. His research interests are solar and stellar astronomy, particularly theories of stellar coronas and winds, and plasma astrophysics, including turbulent plasma heating and transport processes.

Harvey Tananbaum is an astrophysicist at the Smithsonian Astrophysical Observatory and a research associate of the Harvard College Observatory. Tananbaum did his undergraduate work at Yale University and received the Ph.D. from the Massachusetts Institute of Technology in 1968. His research interests include x-ray astronomy, quasars, and active galactic nuclei. He is a contributor to the standard reference work *X-Ray Astronomy* (D. Reidel, 1974), and he served as principal scientist for the Smithsonian's experiment aboard the Einstein (HEAO-2) X-Ray Satellite Observatory.

Robert F. C. Vessot is a physicist at the Smithsonian Astrophysical Observatory and a research associate of the Harvard College Observatory. He did both his undergraduate and graduate work at McGill University in Montreal, receiving the Ph.D. in 1957. His research interests include the development of atomic clocks, microwave transmission systems, and experimental gravitation and relativity tests. In 1978, he received the NASA award for Exceptional Scientific Achievement.

Fred L. Whipple is senior scientist and director emeritus of the Smithsonian Astrophysical Observatory and Phillips Professor of Astronomy Emeritus at Harvard University. Whipple received the A.B. from the University of California at Los Angeles in 1927 and the Ph.D. from the University of California at Berkeley in 1931. Best known for his pioneering work in comets and meteors, his research interests have also included solar system evolution and geodynamics. A prolific writer for both technical and popular audiences, he is the author of *Earth, Moon, and Planets* (Harvard University Press, 1968) and *Orbiting the Sun: Planets and Satellites of the Solar System* (Harvard University Press, 1981) and coauthor of *Survey of the Universe* (Prentice-Hall, 1970).

Index